Norman C. Lee

Understanding Blow Molding

Norman C. Lee

Understanding
Blow Molding

2nd Edition

HANSER

Hanser Publishers, Munich • Hanser Gardner Publications, Cincinnati

The Author:
Norman Lee, 2705 New Garden Road East, Greensboro, N.C. 27455-2815 USA

Distributed in the USA and in Canada by
Hanser Gardner Publications, Inc.
6915 Valley Avenue, Cincinnati, Ohio 45244-3029, USA
Fax: (513) 527-8801
Phone: (513) 527-8977 or 1-800-950-8977
www.hansergardner.com

Distributed in all other countries by
Carl Hanser Verlag
Postfach 86 04 20, 81631 München, Germany
Fax: +49 (89) 98 48 09
www.hanser.de

Library of Congress Cataloging-in-Publication Data
Lee, Norman C., 1934-
Understanding blow molding / Norman C. Lee. -- 2nd ed.
p. cm.
ISBN-13: 978-1-56990-416-9 (flexicover)
ISBN-10: 1-56990-416-2 (flexicover)
1. Blow molding. I. Title. II. Title: Blow molding.
TP1150.L4 2000
668.4'12--dc22
2007014471

Bibliografische Information Der Deutschen Bibliothek:
Die Deutsche Bibliothek verzeichnet diese Publikation in der Deutschen Nationalbibliografie;
detaillierte bibliografische Daten sind im Internet über <http://dnb.d-nb.de> abrufbar.

ISBN: 978-3-446-41265-1

Production Management: Oswald Immel
Typeset by PTP-Berlin, Berlin, Germany
Coverconcept: Marc Müller-Bremer, Rebranding, München, Germany
Coverdesign: MCP • Susanne Kraus GbR, Holzkirchen, Germany
Printed by Kösel, Krugzell, Germany

Introduction to the Series

In order to keep up in today's world of rapidly changing technology we need to open our eyes and ears and, most importantly, our minds to new scientific ideas and methods, new engineering approaches and manufacturing technologies and new product design and applications. As students graduate from college and either pursue academic polymer research or start their careers in the plastics industry, they are exposed to problems, materials, instruments and machines that are unfamiliar to them. Similarly, many working scientists and engineers who change jobs must quickly get up to speed in their new environment.

To satisfy the needs of these "newcomers" to various fields of polymer science and plastics engineering, we have invited a number of scientists and engineers, who are experts in their field and also good communicators, to write short, introductory books which let the reader "understand" the topic rather than to overwhelm him/her with a mass of facts and data. We have encouraged our authors to write the kind of book that can be read profitably by a beginner, such as a new company employee or a student, but also by someone familiar with the subject, who will gain new insights and a new perspective.

Over the years this series of **Understanding** books will provide a library of mini-tutorials on a variety of fundamental as well as technical subjects. Each book will serve as a rapid entry point or "short course" to a particular subject, and we sincerely hope that the readers will reap immediate benefits when applying this knowledge to their research or work-related problems.

E.H. Immergut

Series Editor

Preface

The material for this book is drawn from a seminar team, lead by the author, that conducts for the Society of Plastic Engineers (S.P.E.) blow molding seminars and workshops on both industrial and high production blow molding (bottles). The focus of *Understanding Blow Molding* is hands on and practical application, which will benefit those new to the plastic blow molding specialty, as well as those who are experienced but who may not have been exposed to all facets of a blow molding plant. People from various disciplines such as marketing, design, research and development, product and manufacturing engineering, as well as operation personnel, will also gain insight into the everyday problems of a blow molding operation.

Those who participated on the seminar team and whose material is quoted and referenced include:

- *John Hsu*, Milacron, Cincinnati, OH (Chapter 7)
- *Robert Slawska*, Proven Technology, Inc., Belle Head, NJ (Chapter 7)
- *John Griep*, Portage Cast & Mold, Portage, WI (Chapter 9)
- *Ralph Abramo*, Abramo Associates, Inc., Holliston, MA (Chapter 10)
- *John McNamara*, Blow Molding Automation Specialist Incorporated, Cincinnati, OH (Chapter 11)
- *Mark McCullon*, Plastic Omnium Zarn, Inc., Reedsville, N. C. (Chapter 11)
- *Ronald Walling*, Advanced Material Center, Ottawah, IL (Chapter 3)

Further, I am grateful for the work of *Loretta A. Lee* of CADWORKS for her assistance in editing, preparing the drawings, and bringing together the pieces of the manuscript, without which this book would not have been possible. Further thanks go to *Aaron Ulmer*, Teaching Assistant with the Technology Department at B.Y.U, for his help with metric conversions.

Norman C. Lee

February 2000

Contents

1 Introduction to Plastic Blow Molding

1.1 Scope of this Book

The concern of this book is with the process of blow molding of thermoplastic materials and to familiarize those who have not been exposed to this technology, whether they are working as a technician, designer, engineer, or salesperson, or if they are a student studying plastics or a related field. This work is not a complete or detailed account of all the areas of the field but is intended to point out some of the various aspects encountered on a daily basis.

To gather more indepth information it is recommended that further study of various books, magazines and videos, etc. [2, 3] be undertaken.

1.2 Background of Plastics

Background

For those who have not been exposed to the plastic industry previously, some background of plastics and plastic technology to the present time follows. The word *plastic* is derived from the Greek word for moldable and pliable. Today plastics almost always denote organic polymers, and this includes colorants, lubricants, plasticizers, etc. *Kunstharz* or *Kunststoffe* are German terms for synthetic resin, which is translated as artificial (synthetic) material. Other materials such as glass, concrete, and steel are also artificial materials. However, because they are inorganic, they are not grouped with plastics [1].

1.2.1 History of Plastics

In the last decade the field of plastics was challenged and depicted as the environmental bad guy. It was pointed out to the general public by the plastics industry that in fact, without plastics, much of the world's industry would come to a standstill. The electrical industry would be without insulators, the rolling mills without lubricants, and paper without waterproofing. The public would miss most of the materials that now provide shelter, preserve food, and provide clothing. Not forgetting that in the information age, which we are now entering, key components would not be possible without plastic.

The demands for natural materials such as ivory, silk, and rubber caused by the industrial revolution, at the turn of the twentieth century, created the environment for new materials to be developed. Chemical engineers seized upon the opportunity to create materials to replace existing hard to obtain and expensive exotic materials with those of lower cost and better quality, and which were more easily produced.

Further development led to successful applications to replace zinc, aluminum, and steel drawn by the need for lighter weight materials and lower production costs. Later and still today, more new materials are being generated to meet the demands of the space age and communication environment.

1.2.2 Polymers

Polymer resins are formed by a variety of chemical processes generally referred to as polymerization. This means that one or more types of molecules are combined to form larger, more complex molecules.

Hydrocarbons

All of the materials (polymers) used in blow molding today are compounds processed and developed from the class of materials generally referred to as hydrocarbons. By way of explanation, typical hydrocarbons, such as polyethylene, can join upon application of heat in the presence of special catalysts. A catalyst controls the way in which the polymer grows and thus its final bulk properties. At high temperatures the chains of polyethylene are long coils. When melted, they move very rapidly; when cooled, they lose motion and become "glassy." At this point they are clear. In a way this is not unlike an old-fashioned carpenter's rule. The chains start folding back on themselves and the polymer becomes crystalline. The previous clarity is lost and the material now becomes opaque. In blow molding or injection molding, the melting occurs in feed or compression sections of the screw pushing the polymer into the mold, or through a die. As the polymer cools in the mold, the crystalline domains that existed in the pellets or flakes being molded reform as the molded part cools. The change in volume that occurs during recrystallization accounts for the mold shrinkage of these materials. Crystallization is influenced by the cooling time, melt temperature, and other factors.

1.2.3 Fundamental Properties

The following can summarize the physical performance of polymers:

1. Average molecular weight
2. Morphology
3. Chain linking
4. Additives, fillers, and reinforcing agents

For those concerned with characterization, a series of tests has been developed as a means to pick an optimal material for a particular application. The plastic material suppliers publish these characteristics.

Answers to the following questions should be determined:

1. What are the characteristics or properties of this material?
2. What role will these characteristics play in how the parts are molded?
3. Will these characteristics make the material suitable for the application being considered?

The properties and selection of materials for blow molding will be examined later in Chapter 3.

1.3 Plastic Polymer Types

It should be understood there are two basic groups of polymer materials, thermosets and thermoplastics.

1.3.1 Thermosets

Typical thermosets

Thermosets use the application of pressure and heat to cause a chemical change. Thus, the finished product is chemically different from the raw material, and it takes a permanent "set," meaning that a thermoset cannot be returned to its original state by melting. In the thermoset family is included: phenolics, also called bakelites, after the inventor of phenolic, ureas, melanesias, alkyds, artificial rubbers, some glues, and some paints.

1.3.2 Thermoplastics

Typical thermoplastic

Thermoplastics have basically the same chemistry before and after processing. They may be reprocessed by grinding them up and remelting, with or without mixing them with virgin materials. Losses do occur due to original processing and reprocessing. Typical thermoplastic materials are polystyrene (PS), nylon, polyethylene (PE), polypropylene (PP), polyvinylchloride (PVC), polyethylene terephialate (PET), and engineering plastics (see Chapter 3). A variety of blow molded products can be seen in Fig. 1.1.

Fig. 1.1: Assortment of blow molded products (Courtesy of Kautex Machines, Inc. North Branch, N.J.)

1.4 References

1. Rees, Herbert, *Understanding Injection Molding Technology*. (1994) Hanser Publishers, Munich, Vienna, New York.
2. Lee, N. C., *Blow Molding Design Guide*. (1998) Hanser Publishers, Munich, Vienna, New York
3. Lee, N. C., *Plastic Blow Molding Handbook*. (1990) Van Nostrand, NY

Recommended Reading

Encyclopedia of Polymer Science Engineering, 2nd Ed. Vol. 2. (1992) John Wiley and Sons, New York

Kaufmann, M., *The First Century of Plastics*. The Plastics & Rubber Institute, (1986) London UK

2 Basic Process

2.1 The Principles of Blow Molding

A simple explanation of the principle of blow molding is a balloon. If you blow air into a plastic tube that is closed on all sides except the point at which the air enters, the tube will expand and take the shape of the mold that is around the tube. As in the case of the balloon, the further the tube is stretched the thinner it gets (Fig. 2.1).

Fig. 2.1: Balloons

Parison or pre-form

The process begins with a plastic resin hot tube called a parison or pre-form. The parison is placed within a split mold with a hollow cavity. The mold sides are then clamped together, pinching and sealing the parison tube. Air is blown into the tube, which expands the hot resin wall into the shape of the cavity; the mold is cooled with water solidifying the resin into the shape of the part. Once cooled, the part is ejected from the mold and trimmed (Fig. 2.2).

Basic process

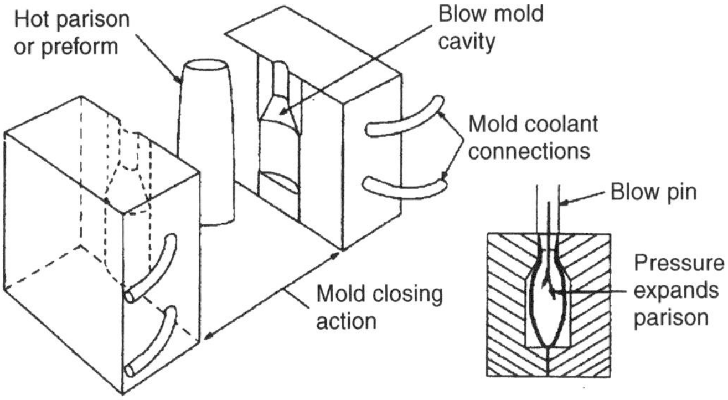

Fig. 2.2: Basic process

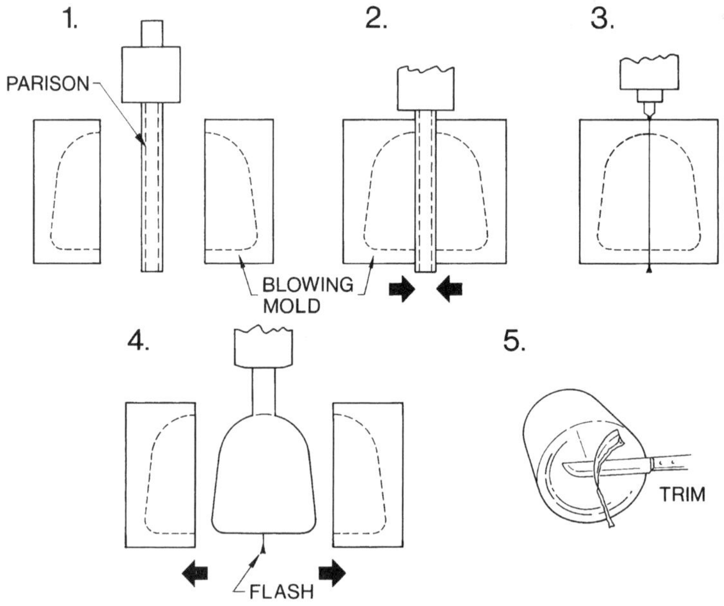

Fig. 2.3: Blow molding stages

There are several methods of blow molding plastic parts. However they all have 5 stages in common (Fig. 2.3).

Five stages of blow molding

1. Platicizing or melting the resin
2. Parison or pre-form production
3. Inflation of the parison or pre-form in a mold to produce the end part
4. Ejection of the part
5. Trimming and finishing of the part

The first four steps take place in sequence; the fifth is performed while the other 4 steps are cycling. Exceptions occur when a number of pieces are produced on the same machine, simultaneously, then the four steps may overlap. Most of the cycle time is taken up by the blowing and cooling step. Therefore (step 3) blowing and cooling control the machine cycle. The speed of the machine that melts the resin and makes the parison must be configured to conform to the blowing/cooling time.

2.2 Parison Formation

The parison or pre-form is formed by either of two techniques for melting the resin, extrusion or injection molding (Fig. 2.4). This diagram shows the family of today's variation of the blow molding process. Each of the processes to be described has its advantages. The end product and volume to be produced determines the best method for the application.

Various blow-molding methods

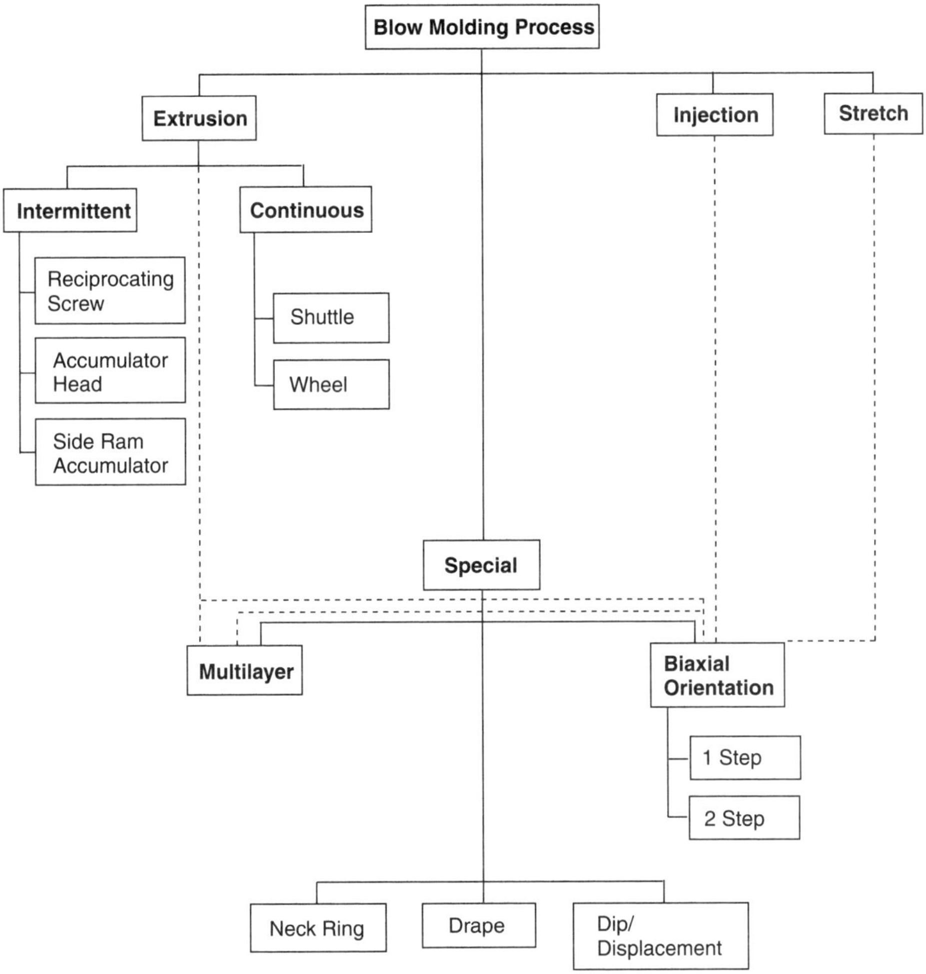

Fig. 2.4: Flow chart

For small, high production clear parts, injection blow molding would be the process of choice. For larger industrial parts the accumulator method is the process of choice; and higher volume detergent or oil bottles would most likely be produced on a wheel machine.

2.3 Blow Molding Development

History

- The first patent issued in the 1850s for blow molding with a material other than glass was issued to Samuel Armstrong. These early items were made from natural latex, mainly novelty items that were unique because of their soft feel.
- The next major advancement came in 1869 with the commercialization of celluloid, which is considered to be the first true thermoplastic material.
- In the 1880s, cellulose nitrate was introduced and was used to produce novelties and toys. This material was softened by steam; the disadvantage was its high volatility, which kept it from being widely used.
- In 1919 a material which was much more stable, cellulose acetate, was available and had much greater acceptance as a commercial material. By 1930 it was available as a squeeze bottle.
- In the early 1930s, Plax Corporation developed the first blow molding machine. This was crude and produced only small quantities. However, from this early beginning a machine to make 25,000 bottles a day was developed. A photograph of the first blow molding machine can be seen in Fig. 2.5.

Fig. 2.5: First blow molding machine. One of the first automatic blow molding machines developed by Plax in the early 1930s. Courtesy of Innopak Corp.

- In 1939 low density polyethylene was introduced by I. C. I. (England). This material was much more amenable to the blow molding process and opened up the way to further development.
- In the 1940s Plax introduced the first LDPE bottle.
- The first injection blow molding machine was introduced. This hybrid machine featured high precision neck dimensions with tolerances as tight as glass bottles.
- Continental Can was issued a patent for a continuous extrusion blow molder in 1950.
- The beginning of industrial blow molding can be said to have started in 1953 with the development of high density polyethylene. This material was developed simultaneously by Phillips Petroleum Co. in the United States and by Professor K. Ziegler in Germany.
- By 1956 the commercialization of the blow molding process and explosive growth of the industry had begun.
- During the 1950s industrial blow molding machines were developed first in Europe, since Plax and Continental tightly held patents in the United States.
- European machines became available in the United States by 1958. The first machine was bought by Empire Plastics to make toy bowling pins.
- Several individuals and companies built machines, and I was personally involved in converting a Reed injection machine to make a blown baseball bat.
- Ideal Toy ordered six machines built by Walden-Hartig (Hartig machines are now made by Davis Standard), to make doll bodies with a head design from Empire Toys.
- By 1960 with patents now freed ZARN, Inc. made milk bottles on Uniloy machines for Borden Dairy in High Point, North Carolina (Fig. 2.6).
- In the '70s biaxially oriented polyethylene terephthalate (PET) was developed, which led to the introduction of the twostep process by which pre-forms and final bottles are produced on separate machines by Cincinnati Milacon, U.S.A. in 1977 Nisser, ASB Company (Japan), began to offer biaxial orientation of PET using blow molding equipment based on a one-step process.
- Also during the '70s, Multilayer blow molding came to the U.S.A with the introduction of the ketchup squeeze bottle.

With the introduction and the application of the microprocessor, resins with a wide range of material properties became available. Also, the availability of larger, more robust equipment and microprocessor technology made the production of a range of industrial products such as automotive fuel tanks, etc., possible. Then too, from Japan and Germany, complex shapes and with irregular contours were possible with the introduction of 3-D blow molding.

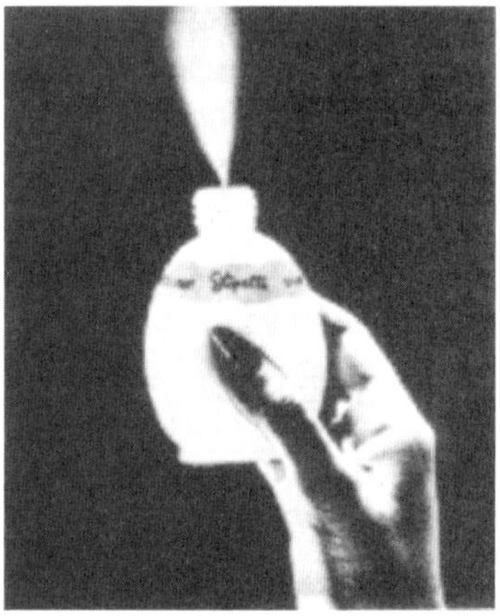

Fig. 2.6: Cellulose acetate (squeeze bottle). First high-volume LDPE blow molded bottle by Plax Corp.

Today bottles still represent a large share of the blow molding market with an 80 percent share; the other 20 percent are considered to be industrial blow molding.

Additional Reading

Whelum, T., *The Bekum Blow Molding Handbook*, Company literature

Lee, N. C., *Blow Molding Design Guide* Chapter 1. (1998) Hanser Publishers, Munich

Mooney, P., Plastic Custom Research Service. Advance, NC

Rosato, D. (Ed), *Blow Molding Handbook*, 2nd ed., (2003) Hanser Publishers, Munich

3 Overview of Polymer Materials

This chapter covers what you always wanted to know about plastics but were afraid to ask!

3.1 Polymer Principles

The term *polymer* is a combination of the prefix *poly*, meaning many, and the suffix *mer*, which means unit. The combination loosely translates to *many units*. The term *plastic* is derived from the description of a type of mechanical deformation under a load. More properly, it would be said plastic is a term applied to a material that will undergo plastic deformation.

Because many polymers will plastically deform, the term plastic was applied and stuck. Many polymers do not plastically deform. Some examples are rubber, certain silicones, thermoplastic rubbers, and many urethanes.

These mechanical properties result from the structure of the polymer's building blocks, which are the monomers. Monomers are single units, utilized in a particular polymer composition.

A few useful terms at this point include:

Polymer building blocks

- Monomer – single unit or basic building block
- Dimer – two or double unit building block
- Trimer – three or triple unit building block
- Polymer – multiple unit building block

Connection in a repeated fashion of monomers is referred to as polymerization or production of the plastic. By this, different combinations of the monomers can be linked, to varying degrees, to each other or to a different monomer, or combination of different monomers.

3.2 Types of Polymers

This section was referred to in Chapter 1. The most common division of polymers is first by the method of processing chemistry. These are commonly referred to as thermoset and thermoplastic (Fig. 3.1).

Thermosets are the class of polymers that undergo a chemical change when pressure and heat are exerted on them resulting in a unique structure different from the monomer and polymer combinations of which they are composed.

Thermoplastics are the class of polymers that do not undergo a chemical change and can be cycled through pressure and temperature cycles to give different shapes which they retain upon cooling to ambient conditions.

Fig. 3.1: Thermoset and thermoplastic

The classes of thermoplastics and thermosets overlap in a gray area. An example of this is cross-linked polyethylene. Other examples are polysiloxanes or silicone resins, also called elastomers.

Blow molding finds the class of thermoplastics to be the most promising for processing.

Thermoplastics are made from monomers that create long chains of the monomer linked together much like boxcars on a train.

Homopolymers

Homopolymers are a special case of polymer in which only one monomer is used. Today, true homopolymers are becoming the exception, where only a decade ago they were predominant in commercial volume. The trend today is to utilize catalysts to put together engineered versions of a polymer.

The development of olefins such as polyethylene, polypropylene, and polybutene which were originally known as oil formers, has drastically changed since the 1970s. Originally the work by the German chemist Ziegler and the Italian chemist Guilo Natta commercialized the manufacture of olefins by the use of catalysts. There are referred to as Ziegler-Natta stereo-specific catalysts.

These small, molecular particles were coated onto a bed of nonreactive larger particles and gases such as ethylene, propylene, and butylene were heated and introduced to the catalyst. This process opened the chemical bonds and formed the monomers into polymers with length and structure to order.

H_2-C = C-H_2	–>	H_2-C-C-H_2
Ethylene Monomer		Free Radical Form

This is achieved by breaking the double bond of the ethylene (or other monomer) and linking it to another by a free radical process. The result is a continuous chain or linking of thousands of these monomer units into a giant polymer. The catalyst allowed control of the process and of the ultimate length of the polyethylene (Fig. 3.2).

Classification of polymers by configuration

Ziegler, Natta, and other scientists identified three types of polymer configuration; these are atactic, isotactic, and syndiotactic. These unfamiliar names indicate the amount of order present in a polymer and represent the first attempt to classify this important characteristic of plastics. Atactic: Very random arrangement, disorderly. Atactic polypropylene is cheeselike. Isotactic: Regular alternation or order. Syndiotactic: Controlled or specific alternation.

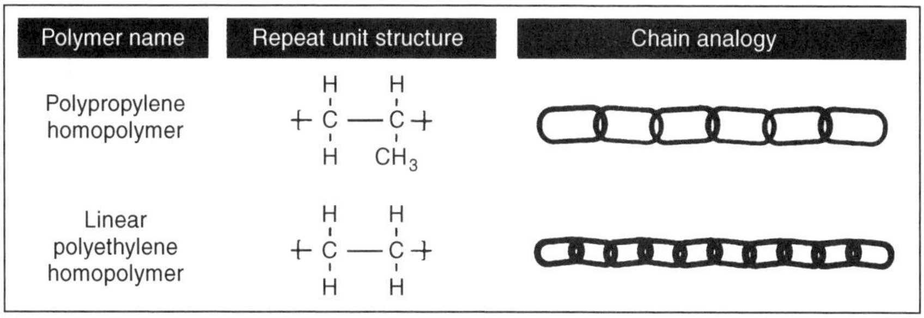

Fig. 3.2: Chain analogy

Influence of molecular weight on properties

Today it is well known that the molecular weight of a polymer has a major influence on its properties. Simply put, the longer the chain, the higher the molecular weight. This is restricted in a practical sense by the fact that the energy needed to extend different monomers to longer chains is the approach of the polymer temperature to its thermal degradation point. That is, it burns. So bulk polymerization is limited by this temperature limit and many novel ways of getting around this limit, such as catalysts, are used.

3.3 Types of Polymerization

There are four main types of polymerization.

Addition: This is simply joining of different monomers together through an opening and subsequent joining of them to form the polymer. An example is: HCI + ethylene = vinyl chloride.

Condensation: This combines two chemical or functional groups on different monomers to form a new unique chemical bond and exclude water or a similar molecule. An example is: glycol + diacid = polyester + water.

Polymerization methods

Free radical: Monomers or polymers and monomers can be coreacted by opening double or triple bonds between carbons. The free radical then essentially moves along the chain and eventually is neutralized or the catalyst is reformed stopping the process. An example is: ethylene (double bond) gets opened by heat, UV radiation, or if a free radical attacks it. The bond now has split into a single connection between carbon atoms and an active site on the carbon atom surfaces. This active site can interact with other similar sites and reform into a single carbon-carbon bond. We have now dimerized or trimerized the first molecule. In fact this will continue along until the motion of the molecules inhibits the process or until a free radical trap stops the process.

Catalytic induced: With the introduction of more complex molecules such as Bucky balls, metallocenes, and new variants of older Ti-Pt and other metal complexes, the conventional polymerization process has rapidly changed and left old workhorse polymers such as conventional high pressure LDPE behind. The Unipol process (originated with the Union Carbide company) was embraced by plastic manufacturers; variants of it have made great strides in production LLDPE (linear low-density polyethylene). Although this is a gas phase process, liquid process

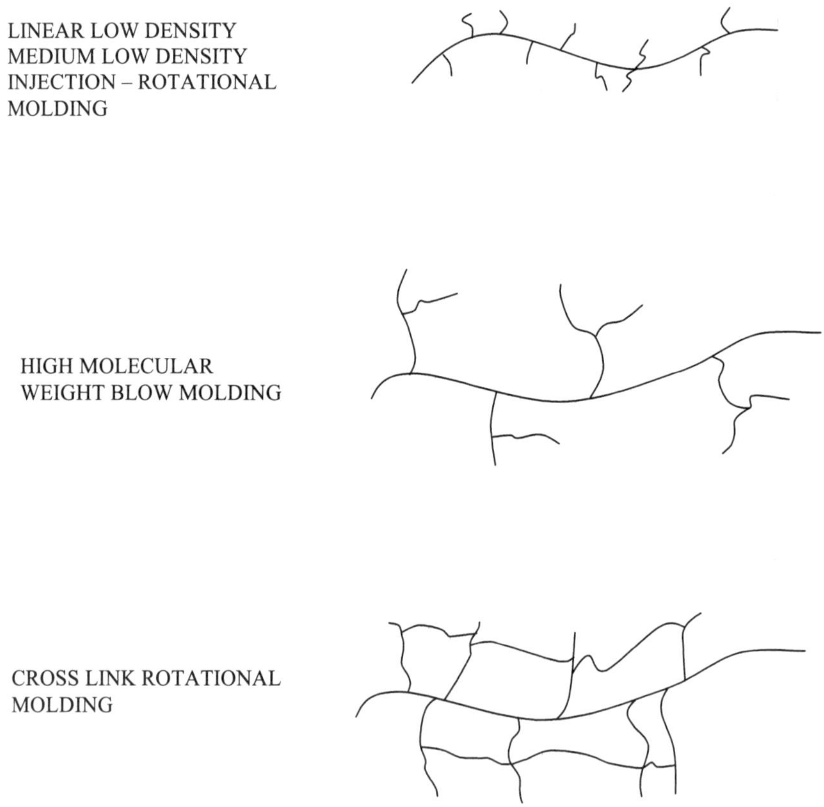

Fig. 3.3: Molecular structure of polyethylenes

Enhancing properties by different polymerization method

work by Dow has resulted in octenebased LLDPE products. An illustration of this can be seen in Fig. 3.3. The early Unipol work used hexene and butene as commoners with ethylene. These have now risen to a new level of properties with the metallocene catalyzed polyethylene now being produced. Some of these are carbon dioxide or carbonyl coreacted to give ECO type resins of superior properties.

This is interesting because these new technologies allow for greater molecular interaction in the polyethylene chains giving better strength and high temperature flow properties, which are needed to blow mold and perform well in large containers. Early attempts to produce superior polyethylene led to the development of UHMWPE. This stands for ultrahigh molecular weight polyethylene. Hoechst in Germany was one pioneer of this. The early materials made were not truly moldable. The material had to be sintered into a product or profile and often finished by machining to make the part. The properties were very good and a large volume of UHMWPE was made. This was one of the first fractional melt materials.

For large containers or shapes the branching of the long chains that was achieved for polyethylene gave low flow (or low MI, melt index) and great strength.

Because a large part parison weighs a lot, the high strength given by the branching allows the HDPE to perform well. Today, branching has given flow rates above the necessary rates to form parts for the food industry.

3.4 Resin Considerations for Blow Molding

3.4.1 Chemical End-Use Requirement

HDPE is impervious to water, water vapor, inorganic chemicals, and inorganic materials in aqueous media, especially at the ambient temperature of normal storage and transport. Typical materials in these categories are:

HDPE resistance

- Most inorganic acids
- Most bases (alkalines)
- Minerals and inorganic chemicals as dry solids and powders
- Minerals and inorganic chemicals in solution or aqueous suspension

Moisture resistance makes HDPE an ideal container material, keeping moisture out as well as in.

With organic chemicals and fluids, HDPE must be used more selectively. Some liquids pose no major problem under ambient conditions. These include:

- Most alcohols
- Most glycols
- Glycerin
- Most organic acids and their derivatives

HDPE absorption and permeation

For hydrocarbon fluids, oils, and solvents (listed later), absorption and permeation through HDPE become a factor.

- Straight chain hydrocarbons-aliphatics (butane, pentane, hexane, heptane, octane, etc.)
- Other hydrocarbons-aromatics (benzene, toluene, xylene, etc.)
- Chlorinated and oxygenated products of above (chlorobenzene, chlorinated solvents, ethers, ketones, esters)

Most of these compounds are flammable and would not normally be contained in HDPE containers without special consideration. Their use as an ingredient in any product being considered for packaging or storage in HDPE requires proper testing to determine package viability.

HDPE stress cracking

Another aspect of chemical end-use requirements that must be considered involves the property of certain chemicals, in either aqueous or organic media, to induce environmental stress cracking. In general, certain detergents and amphipathic chemicals will stress crack HDPE. To resist this effect, high performance resins, primarily copolymers, have been developed.

3.4.2 Physical and Processing Requirements

The HDPE resins used in blow molding have a high molecular weight (MW) to provide the melt viscosity required for proper extrusion and melt strength in the parison. They also have a broad molecular weight distribution (MWD) to provide suitable flow response to processing pressures.

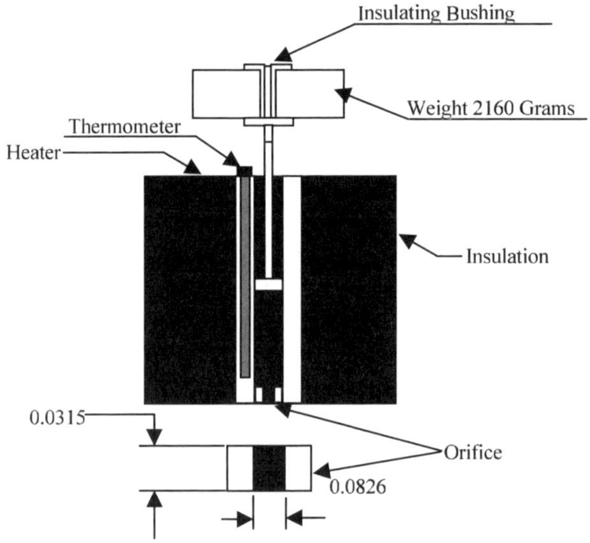

Fig. 3.4: Melt indexer

Melt Viscosity Ratings

HDPE blow molding resins generally have a fractional melt index (MI) of 0.9 units or lower. However, low MI values, indicative of high MW, are misleading when comparing the melt processability of one resin with another. MI measurements are made on a simple laboratory capillary instrument. They represent the weight of extrudate (grams collected in 10 min) from a sample at 190 °C (374 °F) forced through a standard orifice under a static pressure of 0.290 MPa (43.3 lb/in^2). Under these conditions, the shear stress in the melt passing through the orifice is only 0.019 MPa (190,000 dyn/cm^2). Although the 190 °C melt may represent a realistic processing temperature for HDPE, the applied pressure, and resulting melt shear stress, are considerably lower than those encountered in molding (Fig. 3.4).

Melt Index measurements and testing

The previous test conditions are specified by ASTM D 1238, Method E. A diagram of the instrument and its critical dimensions is presented in Fig. 3.4. To provide a better perspective of process fluidity for HDPE blow molding resins, modifications to this procedure are provided by other methods specified in D1238 using the same instrument, at the same melt temperature, but with higher applied static pressures. Table 3.1 lists these methods with the corresponding conditions.

Table 3.1: ASTM D1238[a]

Method	Weight (g)	Static Pressure (MPa)	Shear Stress (Pa)
E	2,160	0.298	19,000
N	10,000	1.38	89,000
F	21,600	2.98	190,000

[a] All at 190 °C

Flow determination via Method N, called melt flow (MF) and Method F, called high load melt index (HLMI) are often provided on specification sheets, particularly for the higher molecular weigh HDPE resins with MI measurements ≤0.1.

Melt Rheometer

Measuring melt fluidity

A melt fluidity measurement can also be made using a gas pressurized rheometer under conditions established by Canadian Industries Ltd. (190 °C, but with an effective pressure of 10.3 MPa (1500 lb/in^2)). The orifice dimensions of this instrument are smaller than those on the melt index device (the diameter is 0.489 mm and the land length is 4.470 mm). The previous pressure induces a melt shear of 0.283×10^6 dyn/cm^2). This rating is called CIL flow index and, like the preceding methods, measures the g of extrudate collected in 10 min.

Basic Melt Rheology

Three commercial HDPE resins with equivalent MIs of 0.2 were rated by the HLMI method and found to vary over a wide range as indicated in Table 3.2.

Table 3.2: Properties of Three Resins

Resin	MI	HLMI	HLMI/MI
X	0.2	32	160
Y	0.2	20	100
Z	0.2	16	80

Such differences in melt properties are of great significances in blow molding performance and equipment design. The higher HLMI/MI ratio indicates greater melt flow under equivalent pressure.

Significance of HLMI/MI ratio for blow molding

Although the MIs indicate equivalent average molecular weights, resin X is more shear sensitive due to a broader molecular weight distribution, especially at the low end of the MWD spectrum. This increased proportion of shorter, linear molecules acts as a lubricant. The ratio of HLMI/MI may thus be used as a guide in rating MWD; the higher ratio always indicates a broader distribution.

The melting and pumping actions of the extrusion process, maintained by heat and pressure, act on the resin via two rheological mechanisms: viscous and elastic. Viscous energy is dissipated through working against the resistance of the system while elastic energy is stored. This stored component causes an overall swell in the extruded parison.

The capacity of a molten HDPE to undergo elastic deformation, and hence store energy, increases as MWD broadens, especially with an increased proportion of larger molecules at the high end of the MWD spectrum.

Melt Swell

The swell effect, as already noted, is induced by the recovery of stored energy in melt elastically. Melt swell, rather than exact die and mandrel dimensions, establishes the final dimensions (both diameter and wall thickness) of the parison. Control of swell, by resin selection or rheology, and swell prediction are essential in precision blow molding operations.

Swell determines final dimension of part

Data on melt swell can be developed using laboratory instruments such as the Melt Indexer (ASTM D1238) and employing different weights in the 2,160 to 21,600 g range. The gas rheometer used in the CIF Flow Index Method, and the Instron Melt Rheometer can provide guidance in predicting melt swell characteristics at temperatures and shear conditions matching those experienced on commercial blow molding equipment. Basic differences in the L/D ratios of the rheometer orifice and equivalent conditions of the parison tooling must be taken into account in developing such comparative data.

The swell ratio (the extrudate diameter divided by the orifice diameter) increases with shear rate. Some resins tend to level off in swell in certain ranges of shear rate, while for others swell continues to increase as shear rate increases. The former types provide better parison dimension control and are preferred. The differences in die configuration between the round orifice of a rheometer and the annulus gap of blow molding tooling are important factors in making comparisons between melt swell and rheological conditions.

Resin Types

HDPE resins used in extrusion blow molding fall into five general categories:

Resins used for blow molding applications

1. Rigid homopolymers (0.960 ± 0.002 g/cm³ range) for thinwall milk, beverage, and food containers. Melt indexes are typically in the 0.7 to 1.00 range.
2. Dual purpose copolymers (0.955 ± 0.002 g/cm³ range) for general purpose chemicals, bleaches, cleaners, automotive products, etc. These provide moderate toughness and environmental stress crack resistance (ESCR). Typical MIs are in the 0.2 to 0.4 range.
3. Detergent grade copolymers (0.950 ± 0.002 g/cm³ range) for special detergents and chemicals where better ESCR is required. Typical MIs are in the 0.3 to 0.5 range.
4. Industrial grade or high performance copolymers (0.943 ± 0.003 g/cm³ range) provide exceptional ESCR and high impact strength. These resins are used for drums up to 76 L (20 gal) and for special containers and structural components. Stable rheological properties are required for the larger parisons. The melt index ratings for these resins are in the 0.1 to 0.2 range and the HLMIs range from 18 to 25.
5. High molecular weight types used for drums up to 210 L (55 gal) and special tanks of larger capacity. Density for these materials range from 0.944 to 0.956 g/cm³. HLMI methods are used to compare flow characteristics; typical values range from 4 to 12.

Homopolymers are preferred for injection blow molding for most pharmaceutical and personal product containers. MIs for these materials are usually in the 0.9 to 1.2 range. Where stress crack resistance is of importance, copolymers are used; these usually have a broad molecular weight distribution for ease of flow in the injection process.

Testing

Environmental stress crack resistance is a most important consideration in selecting HDPE blow molding resins. There are specified procedures for performing an ESCR test, which provide ratings for comparing one resin with another. Resin samples are compression molded to specified thickness, cut to required specimen size, and exposed to chemical reagents under induced stresses. Details for conducting ESCR tests are presented in ASTM procedures

- D-1693 – known as the BTL (Bell Telephone Lab) Method
- D-2552 – known as the CL (constant load) Method

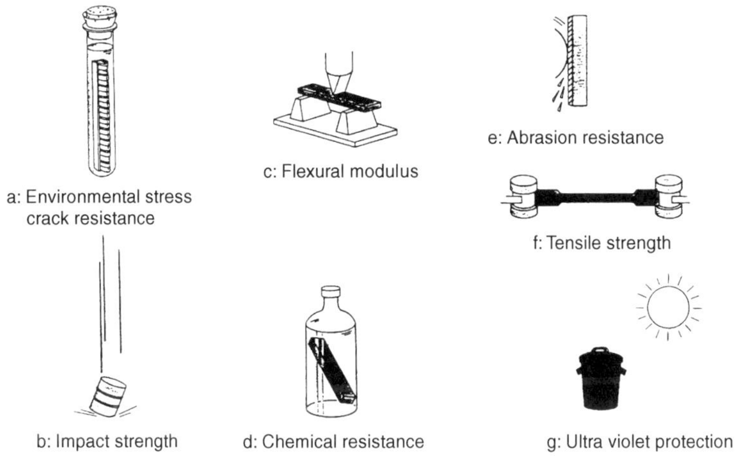

Fig. 3.5: Properties of plastic materials

Testing of Plastics

Number of international standards

The properties of plastic materials are important to the assurance of an adequate, consistent product. These properties are illustrated in Fig. 3.5. The monitoring of the material from its arrival in a molding facility to its departure as a finished item is very important to the manufacturer.

Several groups have been developed to create and promulgate good test practices and methods. In the United States groups such as ASTM and ANSI perform this function. In Canada, CSA does. Japan has JIS and MITI, Germany has DIN, and Great Britain has BIS.

On a more global scale the ISO, the International Standards Organization, is becoming a common ground for the differences in method and practice from nation to nation.

Need for harmonization

The necessity of this harmonization is most evident with the building of plants to manufacture goods by a corporation going from country to country. Everything from the units used to the rate of strain on a plastic part while undergoing a tensile test needs uniformity of design. At the same time there will always be exceptions to rules and test methods.

Testing protocols

These are handled by the subtext in these standards to create the documentation of variation utilized in a particular case. As an example, a blow molding professional tested a material per the ASTM D 695 compression test and ran at a certain temperature to meet the common use of the material. The standard temperature of 23 °C is not a restriction but a common test point. The method allows one to run the test wherever necessary and guides the practitioner to cover all salient points. By following these protocols a designer in Austria can indicate properties for a materials, and a molder in Mexico can identify the grade the supplier has that will qualify to be used. The data is sent electronically and the item is molded. The common tests used on a molded product are listed in Table 3.3. Many more tests are considered important, and this is not meant to indicate others are not. This list is only a starting point.

Table 3.3: Test for Material Properties of Plastics

Test	Designation	Description
Tensile	ASTM D 638	Strength properties of plastics in tension; modulus, elongation, tensile, rupture
Tensile	ASTM D 882	Similar to 638, but for film and sheeting
Compression	ASTM D 695	Compression yield, strength, modulus
Melt index	ASTM D1693	ESCR test for ethylene plastics
Melt index	ASTM D2561	ESCR test for blow molded PE bottles
Hardness	ASTM D2240	Shore A & D hardness of plastics
Hardness	ASTM D2583	Barcol hardness of plastics
Hardness	ASTM D 785	Rockwell hardness of plastics

Additional Reading

Quantum U.S.I. Division. Polyolefin Blow Molding (Equistar Chemicals) and Operation Manual. Company literature

Walling, R., SPE Large Part and Industrial Blow Mold Seminars (Materials Session), SPE

Mallory, R. A., Design Watch, Remember the Four Material Fundamentals, *Plastic World.* (Nov. 1995) (Author's note: Although this work is directed toward injection molding the same basic principles apply here.)

Hartig Training Manual. Davis-Standard Corporation. Edison, NJ. Company literature

4 Extrusion Blow Molding

Extrusion may be subdivided into two methods, continuous extrusion and intermittent blow molding.

4.1 Extrusion

Extrusion is the process of applying heat and pressure to melt the resin and force it through an accurately dimensioned die to produce the desired shape. For blowing purposes this is a shape from which the parison is cut. There are several main parts to an extrusion blow molding machine, which are:

Main parts of extrusion blow molding machine

- Screw
- Barrel
- Hopper
- Feed section
- Compression section
- Metering section
- Screen pack
- Breaker plates
- Adapter
- Die head
- Core
- Mandrel
- Die tip

In the various sections the resin is melted, plastized, and delivered to a die or dies at the proper temperature at a uniform rate (Fig. 4.1).

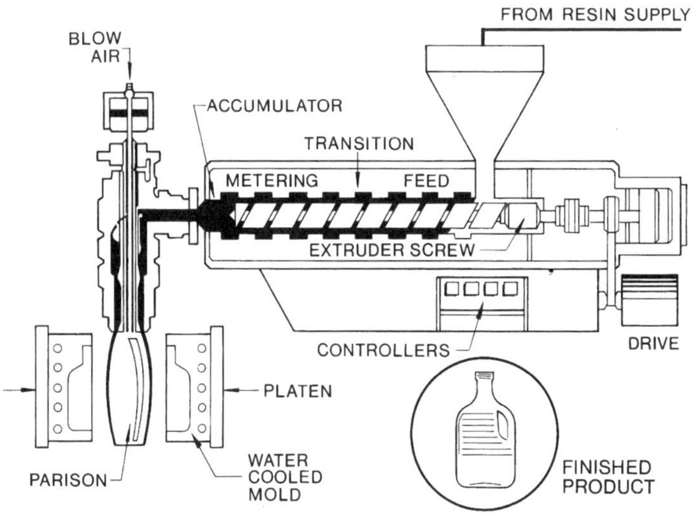

Fig. 4.1: Extrusion machine

Feeding

Extruder sizes are given by bore sizes and the ratio of bore diameter to length of screw. To feed the barrel, resin is first fed into funnelshaped hoppers, which may be done by hand from 22 kg (50 lb) bags or by an automatic vacuum device. The hopper is kept filled with resin. An observation window is on one of its sides to view the resin levels. A cover on the hopper top keeps out contamination from dust, dirt, or other foreign matter, as well as preventing damage to the equipment. The bottom of the hopper has a device so resin flow can be stopped and emptied. The hopper throat and part of the feed section are water or air cooled to prevent resin from melting too soon and sticking together and bridging in the hopper throat before reaching the screw. Bridging will cause problems, such as shutting down production and oxidating of the resin in the barrel, requiring cleaning of the extruder screw and die. From the hopper throat the resin enters the channel of the extruder screw.

Melting and mixing

The screw is driven by a variable speed drive and rotates within the hardened liner of the barrel. The rotating screw flights force the resin forward. As the screw turns the resin is heated, mixed, and compressed inside the screw channels. A long, well engineered screw and barrel make for better melting and mixing of the resin. This provides a better appearance and wall thickness uniformity in the parison. The production rate is also improved, and the melt temperature is easier to control. It is recommended that the screw be at least 20 times longer that its diameter and have a compression ratio of 3.5:1 to 4:1. Since the rotating screw creates frictional heat, excessive heat in the metering section of the barrel may cause resin degradation. Excess heat may be removed with cooling fans, which when operated along with the heat controls maintain a constant temperature. Oil or water channels may also be used to reduce excessive heat. Although considerable frictional heat provides the internally generated melt of the resin, external heating is always required, especially during start-up. Upon reaching the end of the screw channel, the melt passes through the screen pack and thence to the breaker plate and through the adapter die or dies. The screen pack and breaker plate serve as a final filter to eliminate any foreign matter that has passed by the hopper screen. The head, or heads, are designed without sharp corners or edges for the resin to pass over. Heat streaks in the finished parison can result from a poorly designed flow channel that has sharp corners or edges. The adapter and die must be heated from the outside to prevent loss of heat from the melt, thereby maintaining the melt temperature obtained in the extruder.

Heating and cooling

If the melt temperature is too low, the parison will be cloudy or dull in appearance and may show signs of skinning or flakiness. A parison of this type will not be hot enough to enable good weld lines at the pinch-off areas. If the melt temperatures are too high, then the parison will look clear and have glossy areas and may emit smoke. The hot parison will be stringy and tends to stretch easily; the result is thin areas. This will cause extremely thin weld lines.

The function of the extrusion blow molded head and die unit is to:

Parison formation

1. Form the melt into its final shape – the parison
2. Maintain the melt at a constant temperature to meter out the melt at a constant pressure, rate, and desired wall thickness

In the die head unit, the melt flow in blow molding is almost always diverted from the horizontal to the vertical and extruded downward. This unit is the crosshead type in which the melt flow is turned by 90°. There are two kinds of heads used (Fig. 4.2):

1. Center feed
2. Side feed

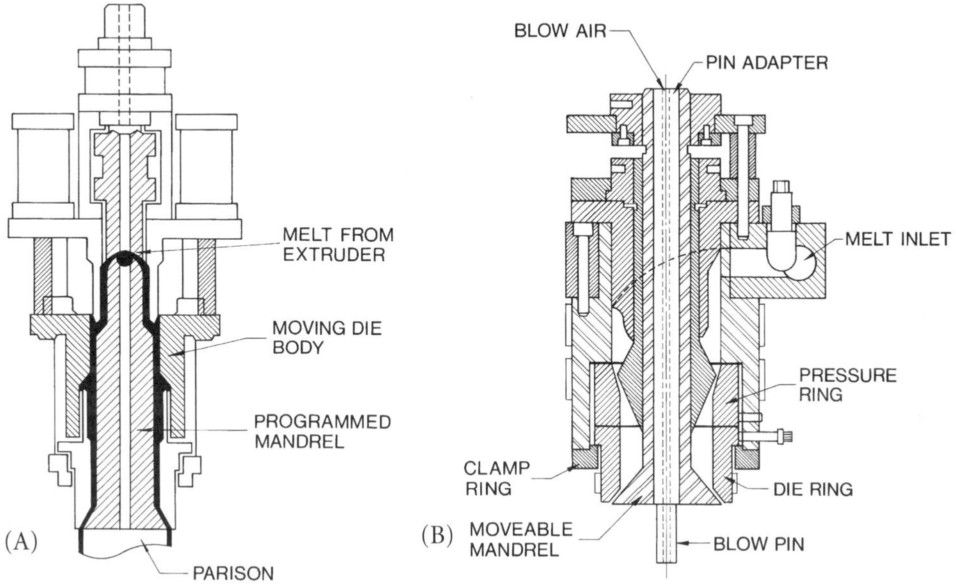

Fig. 4.2: (A) Center feed; (B) Side feed

Types of die heads

Center feed dies have the advantage that the melt core is fairly uniform. The melt flows downward toward the tip of the torpedo-shaped core for a certain distance after having been turned by 90° in the die head, resulting in straight flow all around the core. The disadvantage of center feed dies is that the core must be supported inside the head-die unit. There are two kinds of support for the head-die unit, the perforated support, and the spider support. The perforated is usually preferred because it will keep the melt flow uniform. Both types give rise to flow lines in the blown part. Flow lines will result in a poor appearance of the blown piece and reduce strength along these lines.

Center feed

Side feed

In side feed dies, the melt flow hits the core. To obtain uniform melt flow around the side feed core, and from the die with a straight and uniform tube to be extruded, a groove is frequently machined into the core. Many molders prefer this groove to be circular and horizontal.

Many other designs are available including one in which two heart-shaped grooves guide the bulk of the melt flow downward.

Good die design

A good die design is another means of improving uniformity of melt flow and creates a relatively long die land. At the face, the melt should flow straight down to form a parison with uniform wall thickness. This ringshaped (annular) section at the end of the die is known as the die land. This is the working area of the die and mandrel where the volume of the resin is kept constant. The die length should be neither too short nor too long. The recommended length is 10 to 40 times the die opening or the width of the die slot. A long die land will increase back-pressure on the melt in the extruder head and improve mixing. A short die land in conjunction with a high extrusion rate usually results in outer surface roughness of the parison and blown part. To find the most favorable die land is a combination of experience and trial and error.

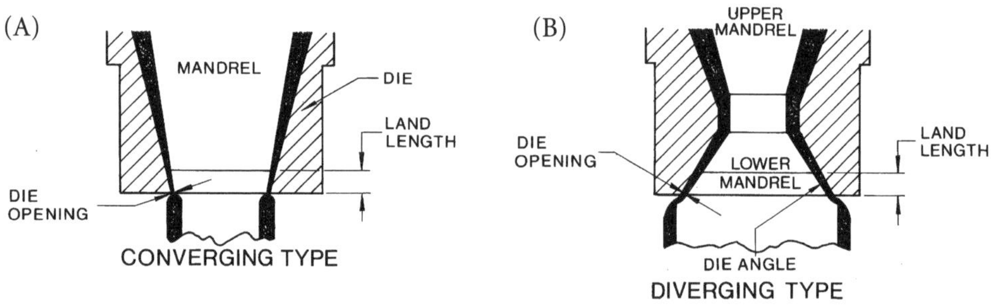

Fig. 4.3: (A) Diverging type; (B) Converging type

4.1.1 Mandrel and Dies

The dies and mandrel are sized to form the parison diameter and wall thickness. There are two shapes of mandrels and dies (Fig. 4.3):

- Converging – which flares inward
- Diverging – which flares outward

Good die design

Typically the hot resin swells as it leaves the die opening or with the die, the slots. This is known as parison die swell (Fig. 4.4). The opening is an accurately machined orifice, through which the melt is pressed to a steady flow. A good die design is internally smooth with no chance of even small particles becoming stuck or held up to overheat or degrade. This results in a solid looking tube coming from the die opening which is made up of a hot fluid, and which has a comparatively high stiffness for a liquid. The parison will stay in one piece unless the temperature is much too high. (This is known as melt strength.)

Influence of gravity

While the tube slowly extrudes to the desired length, it is of course subject to the downward pull of gravity; therefore unless the extrusion is very fast, the wall will thin out at the top and thicken toward the bottom. This is known as sag, draw down, neck down, or stretch out. Thus the final part will end up being thick at the bottom and thin at the top, resulting in an uneven wall thickness. When the parison reaches a critical length it appears to be extruding faster, indicating the upper part is stretching and thinning out. This may be overcome to a degree by gradually increasing the wall thickness during extrusion by means of a movable mandrel[1] inside the die.

4.1.2 Wall Thickness

Uniform wall thickness

The pin[1] in the die may be activated by a timing mechanism called an automatic timing device, which raises or lowers the pin while the parison is being formed. This allows the resin to be extruded to form a parison that will have a more uniform wall thickness, matching the sagging which would otherwise be thin in the top area of the tube. When the mold closes, the pin returns to its original position. It can be seen by moving the mandrel up and down during

1 terms mandrel and pin may be used interchangably

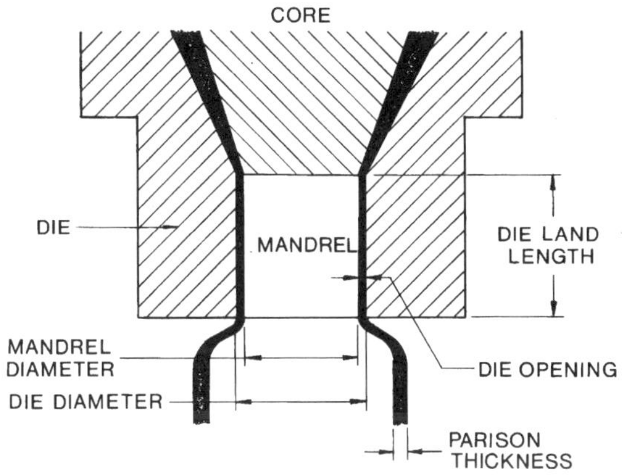

Fig. 4.4: Parison die swell

the extrusion of the parison. Additional resin (and increase of wall thickness) can compensate for irregular part configurations. This is achieved with preset automatic variable die opening and is known as programming (Fig. 4.5).

Die programming

A further method of controlling wall thickness of the parison is varying the extrusion pressures through a fixed die opening. Higher pressure causes an increase in output, while lower pressure will cause less resin to be extruded.

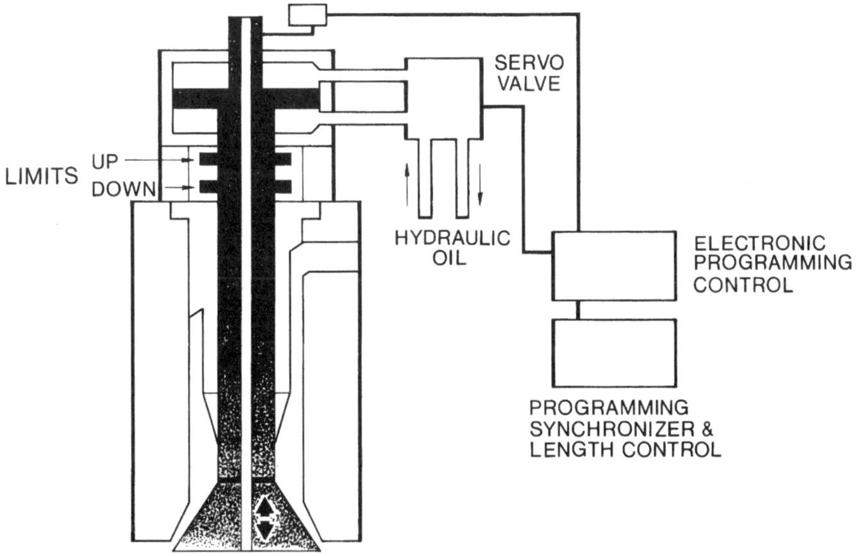

Fig. 4.5: Moving mandrel programming

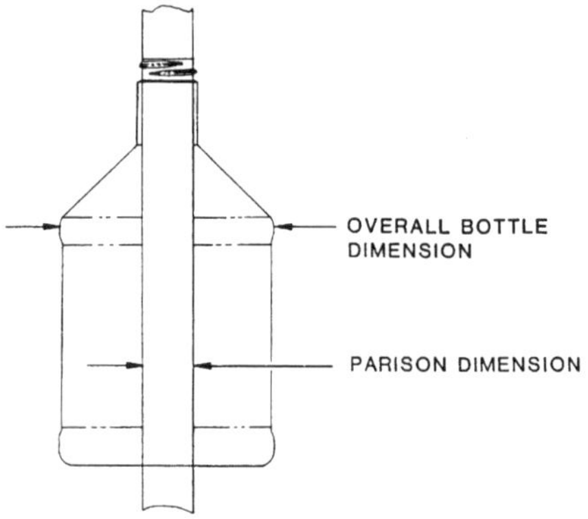

Fig. 4.6: Blow-up ratio

4.1.3 Blow-Up Ratio

Blow-up ratio

Blow-up ratio is defined as the ratio of the average diameter of the finished product to the average diameter of the parison. The maximum recommended blow-up ratio for most applications is 5:1 but 3:1 is preferable and the optimum result also depends on wall thickness. See Fig. 4.6 for an illustrion of the blow-up ratio.

With conventional tooling the mold closing around the parison at the pinch-off causes the parison to change in cross sectional shape (like a toothpaste tube). The result is less blow-up distance at the parting line compared with the distance at 90° from the parting line.

4.1.4 Validation

Providing more uniform wall thickness by ovalization

The working land of the tooling may have clearence increased between die and mandrel in areas where greater blow-up ratios are needed. As a result, the parison will have walls that in local areas are vertical on the circumference, thus providing a more uniform wall in the circumferential direction as well. This is known as ovalization (Fig. 4.7).

Ovalization may be done either on the die or mandrel but is usually on the die since it is easier to maintain proper alignment. Generally the validation should not be completely oval in shape but only where a greater blow-up ratio is required.

4.1.5 Wall Thickness and Melt Temperature

As might be expected, melt temperature becomes critical when extruding a parison with varying wall thickness. Should the melt temperature be too low, the thick areas may end up as thin areas. This can occur because of differences in stretchability and heat retention in the thicker areas.

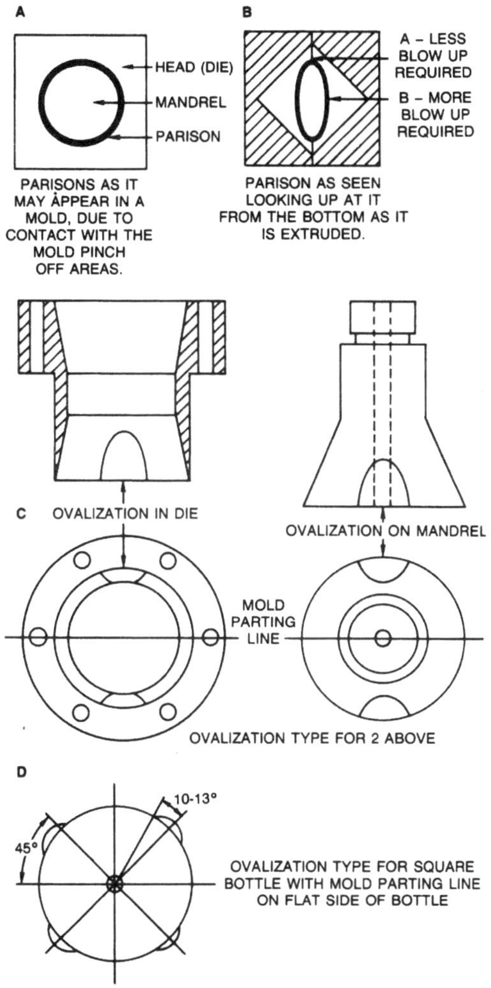

Providing uniform wall thickness by tooling ovalization

Fig. 4.7: Tooling ovalization. (A) Less blow-up required; (B) more blow-up required; (C) ovalization in die (left) and ovalization on mandrel (right); (D) ovalization type for square bottle with mold parting line on flat side of bottle

4.2 Continuous Extrusion

4.2.1 Overview

The extruder produces an endless parison, which subsequently is pinched and/or cut.

In a shuttle or reciprocating mold, frequently of the multicavity type with one or more parisons, the parison is extruded in one stage from one head and die unit. The bottom or side holds a blow pin or needle, which enters the parison while the mold halves are rising or sliding sideways.

Bottle blown in vertical mold

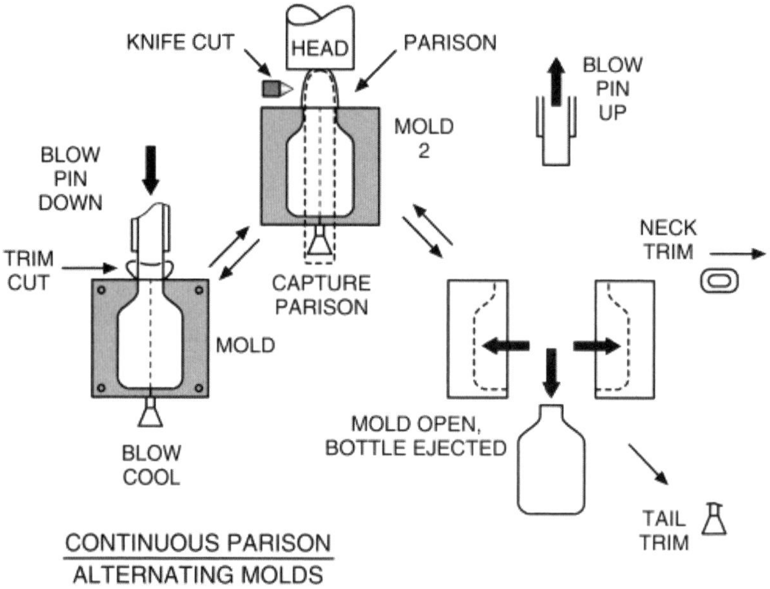

Fig. 4.8: Rising vertical mold

4.2.2 Rising (Vertical Movement) Mold

A hydraulically operated clamp rises and the mold halves mounted on the platens close around the parison, separating it from the extruding tube. Sometimes a cutting device is added to the bottom of the die. Then the clamp drops back down for the blowing and cooling cycle. After ejection of the part, the clamp rises again to separate another parison which has been extruded in the meantime. See Fig. 4.8 for an illustration of the rising vertical mold.

4.2.3 Sliding (Horizontal Movement) Mold

In this piece of equipment the clamp assembly moves to the side (left or right) and a second assembly clamps the mold over the continuously extruding parison. While the first mold is blowing/cooling after the part is ejected, the clamp assembly is ready to move back to pick up another parison (Fig. 4.9).

Shorter cycle time

Both of these methods gain cycle time since no time is lost while the the blowing/cooling is completed and the cycle can be continuously repeated. The parison extrusion rate is adjusted so that it matches the lag time required for the blowing/cooling stage of the cycle.

The clamp assembly with the mold opening and closing must move at a rate exceeding that of the parison extrusion. A further variation of the continuous process is the directional valves at the end of the extruder which divert the stream of melt from one side to the other with a die head at each side. One side is therefore blowing/cooling as the other is extruding the parison. Again the extrusion time has to be matched with the blowing/cooling time. In all the foregoing, the die heads may have several parison streams so that multicavities can produce

Bottles blown on sliding horizontal machine

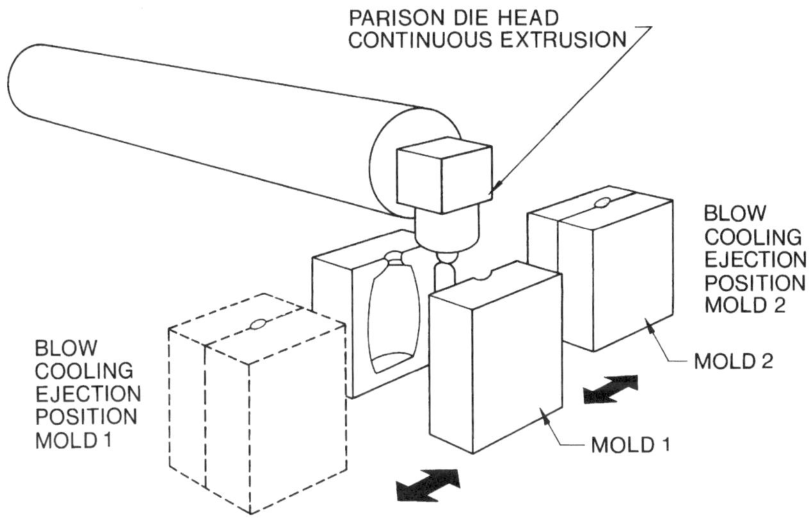

Fig. 4.9: Sliding horizontal moving mold

several parts at the same time. A blow pin needle positioned in the mold, preferably near the pinch-off to penetrate the parison effectively, can affect blowing. In the case of bottle molds, a blow pin is mounted in the clamp assembly.

4.2.4 Molds on a Rotary Wheel

Method for high-volume production

For very high production long run parts, such as personal care products, a rotary wheel is the preferred method. A vertical wheel is mounted next to the extrusion die, on which a series of split blow molds are mounted. Each of the molds is independently operated either hydraulically or pneumatically. While the wheel turns, one mold at a time passes under the die to receive and close around the parison; this motion is continuous. The blowing is introduced through a hollow needle and penetrates the parison when the mold closes or shortly afterwards. The air is fed through the wheel axle as is the water used to cool the molds. Rotary cams, located in the wheel axle, govern the mold opening, mold closing, blowing air start, and blowing needle insertion (Fig. 4.10).

Machines of this type may extrude the parison either downward or upward where it would be pulled along as needed.

Upright rotary wheels may carry up to 20 or more molds. Rotation of the mold carrier is arranged to provide ample time for an open mold to blow, cool, and eject the end product. The extrusion parts are connected by sections of the tube, which require trimming and finishing steps at both ends. The weight, and therefore the wall thickness of the blown part, can be reduced or increased (within limits) by varying the extruder speed or rotation speed of the wheel.

Two types of rotary wheels

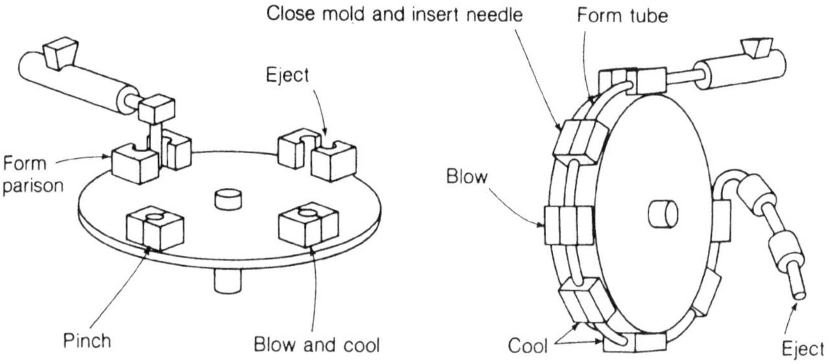

Fig. 4.10: Continuous multi molds

4.3 Intermittent Method

The intermittent method produces a parison by the extrusion process. An accumulator chamber is filled and a hydraulically operated ram pushes the resin through the die and mandrel. This may also be achieved with a reciprocating screw.

Production method for large parts

The intermittent method is usually used to make large industrial parts. An accumulator is a reservoir for gathering and retaining a substantial volume of well-plasticized melt coming from a regular extruder. The accumulator may force melt at high speed out through the die to form the parison tube. The use of an accumulator allows the intermittent delivery rate from the die, and thus the time to form the parison, to be independent of the delivery rate of the extruder. Note that the resin extrusion into the accumulator is slow compared with the fast extrusion of the parison by the ram. The accumulator capacity determines the maximum size of large blown parts. An accumulator may hold 54.43 kg (120 LB) of resin which forms into the parison for a single part. See Fig. 4.11 for an illustration of the intermittent parison method.

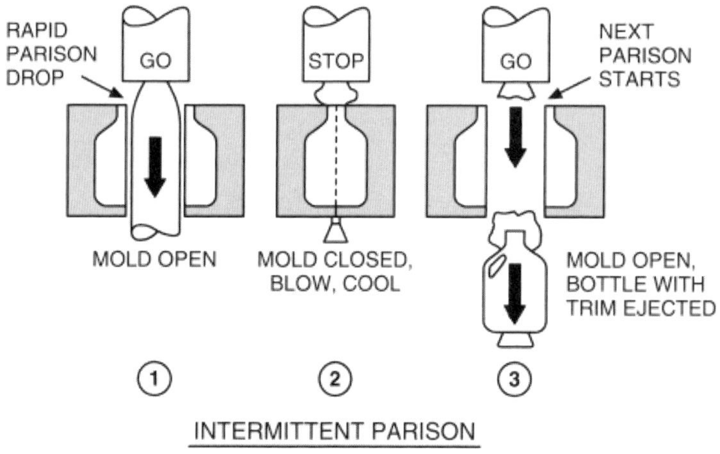

Fig. 4.11: Intermittent parison mold

The advantages of an accumulator are:

Advantages of intermittent method

1. Holds ready a larger volume of melt for large items requiring very long molding (cooling) cycles up to several minutes
2. Permits high production rates, for large parts
3. Enables rapid extrusion of large parisons (short parison formation time) and consequently a short suspension time for the cooling parison and thus comparatively little sag
4. Enables uniform shot sizes and thus a minimum amount of waste
5. Decreases idle mold time to a minimum

Additional Reading

Lee, N. C. (Ed.), *Plastic Blow Molding Handbook.* (1990) Van Nostrand Reinhold, New York

Lee, N. C., *Blow Molding Design Guide* Chapter 8. (1998) Hanser Publishers, Munich

Quantum U.S.I. Division, Polyolefin Blow Molding (Equistar Chemicals) and Operation Manual, Company literature

5 Injection Blow Molding

The injection method of blow molding consists of injecting a thermoplastic material into a cavity and around a core rod producing a hollow test tube like shape, which as previously mentioned, is called a pre-form. The molded pre-form, still on the core rod, is transferred into the blow mold. Clamp tonnage is applied to the mold to withstand the air pressure which blows the hot pre-form into the shape of the cavity. Water lines in the cavity in turn cool the part. This requires a standard injection molding machine and a mold to produce the pre-form, which may be transferred to the blowing mold or stored and reheated. Further variation occurs when the pre-form is stretched. An illustration of injection blow molding is shown in Fig. 5.1.

Producing pre-forms for immediate or subsequent blowing

5.1 Injection Molding Method

The injection machine is used to create a parison (or pre-form). In this process, the pre-form is molded by injecting the plastic material into a split cavity, top and bottom, over a core. The part is cooled and indexed to the next station where split cavities in the shape of the part are closed over the core, and the previously molded pre-form is blown to fill the cavity shape. Next, the cavities are opened and are indexed to the next station for ejection. Sometimes an extra station is used for conditioning. In this injection process cycle, the ram moves rapidly forward and forces the resin into the front of the cylinder, compacting it tightly behind and around the centrally located plasticizing torpedo. Fins keep the torpedo centered. These fins also serve to transfer heat from the hot cylinder walls.

Molding the pre-form

As with extrusion, heat and pressure act together in plasticizing the resin. The melt moves forward in the cylinder. As the melt temperature meets that of the cylinder, it is mixed in the passage around the torpedo and becomes more fluid at the time it reaches the cylinder nozzle. As the resin is melted in the hot cylinder, the pressure from the ram squeezes the air out. The air flows back along the clearance between the ram and barrel and back out through the hopper.

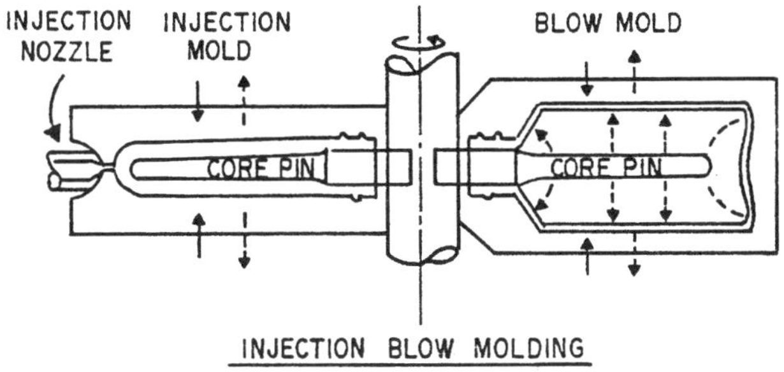

Fig. 5.1: Injection blow molding

The melt is forced by the ram through the nozzle at a very high rate of flow, through the nozzle and sprue into the mold. The sprue may connect directly with the cavity or to narrow channels (runners). Two basic variables must be in balance in the barrel, injection temperature and ram pressure. Since most resins become more fluid by increasing the temperature, the pressure to fill a given mold is dependent on melt temperature. Therefore, the molder has four variables which can be adjusted to obtain pre-form mold fill.

The four variables are:

Process parameters to ensure proper mold fill

1. Temperature
2. Pressure
3. Feed rate
4. Injection time

Minimizing cycle time

Should a particular pre-form be making short shots, increase the temperature to obtain lower resin viscosity; or an increase in pressure, feed rate, or injection time will give the pre-form desired. The same operation conditions are found for molding pre-forms as for any good injection molded product. It is important however to moderate injection temperatures and minimum pressures, and also to warm pre-form molds. While for all blow-molding steps cycle time is important, a short injection time is absolutely essential. However, the barrel temperature must be kept sufficiently low so that no cooling is necessary to reach the temperature at which the material should be blown. Also, the time in the blowing mold must be kept to a minimum so that total cycle time is kept to a minimum.

Disadvantage

Pre-forms may be made in single cavity and multicavity molds, which usually run hot with little or no attempt to cool them. Indeed, often a heating fluid is used to maintain a desired temperature distribution in the mold. As indicated earlier the molding process uses two molds, the pre-form mold and the blow mold. Thus, the expense is greater than for the extrusion blowing method.

An injection mold is more complicated and must be made to withstand injection pressures. Pre-form cooling during transfer creates strains; and marks from the pre-form mold parting line will also appear in the final part.

Advantage

In injection blow molding, the core in the pre-form takes the place of the mandrel in the extrusion process. The pre-form is injection molded around this core. The core may be hollow which can be transferred with the pre-form to the blow mold. It may then be used in the pre-form as the blow pin in which the blow air is introduced. While one pre-form is being blown, the next pre-form is being injected around another core. An advantage of the injection molding process is that the bottle neck or opening finish may be molded in the pre-form mold, before the pre-form is transferred to the blow mold. Also, the blown end product requires no finishing, reaming, smoothing, trimming, etc. Should there be a sprue at the bottom of the pre-form, it is cut off by the closing mold. Further, the part may mold in such a way that the bottom is smooth and unmarked by a sprue and without pinch-off waste.

Irregularly shaped hollow items with uniform wall thickness

As the pre-form hollow shape is injection molded instead of being an extruded tube as in extrusion blow molding, it can be given any nonuniform wall thickness. Thus, because the pre-form is nonuniform, it is possible to obtain irregularly shaped hollow items with nearly uniform wall thickness. The dimensions of the pre-form mold and the pre-form must be carefully and accurately tailored in order to accomplish the desired uniform shape.

There are two methods of transferring the pre-form to the mold:

1. Hand transfer – this method is rarely used in production today and is only found in laboratories for research and development purposes
2. Mechanical transfer (the preferred method)

Types of injection blow molding machines

Two styles of standard injection blow mold machines are available, the three station and the four station (Figs. 5.2 and 5.3). The stations are as follows:

1. The first station consists of the injection station which molds the pre-form
2. The second station is the blowing station with the blow mold
3. The third station is the stripping station
4. The fourth station is an extra position, which acts as a safety station. Here the core rods can be checked for pre-forms or bottles not stripped off prior to indexing to the next station. This station can also be used to condition the outside and inside of the core rods to ensure they are at the correct temperature prior to entering the next station.

Cycle on 3-station injection blow molding machine

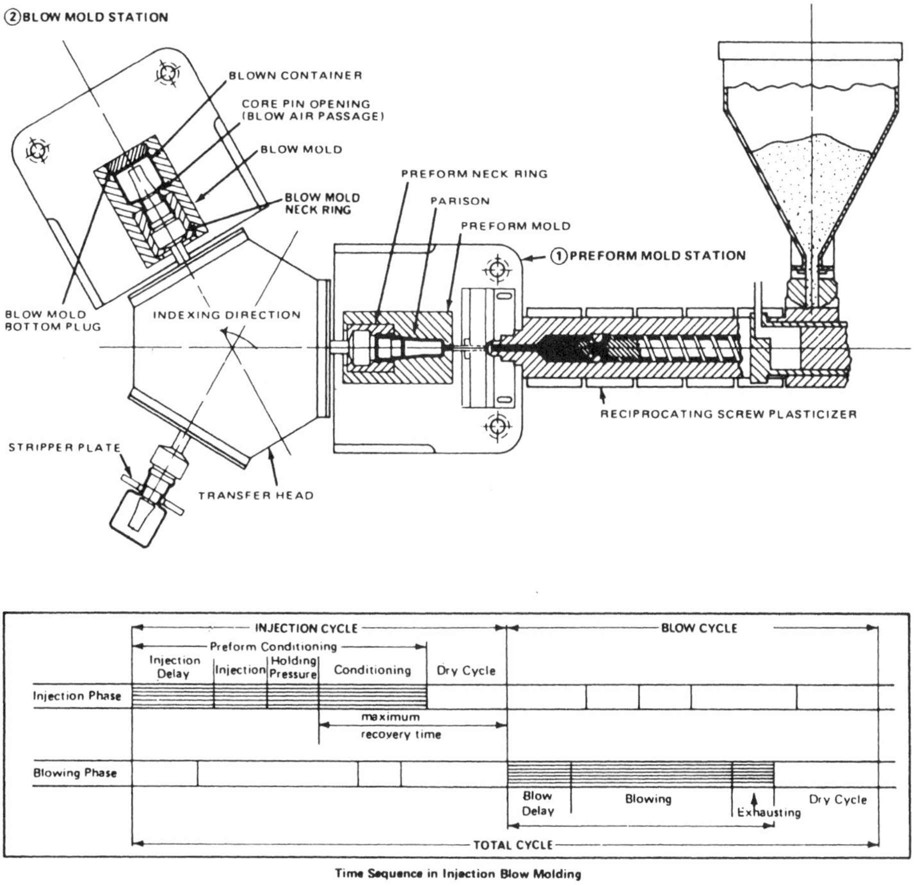

Fig. 5.2: Three station injection blow molding

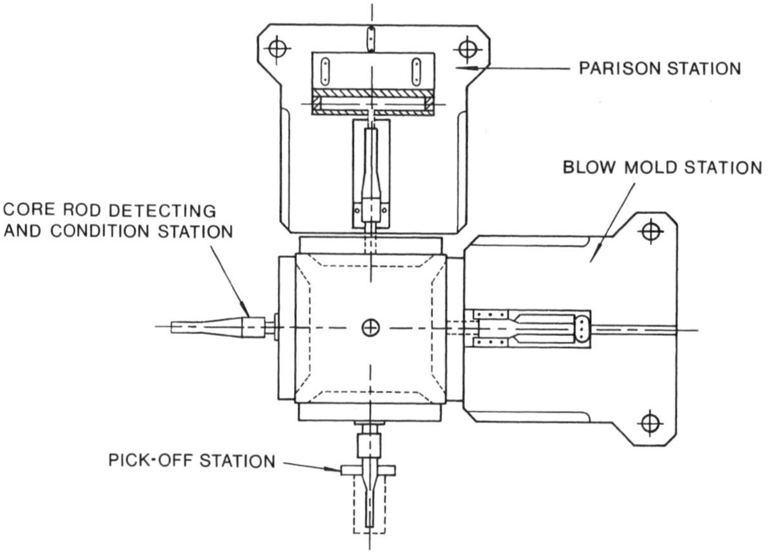

Fig. 5.3: Four station layout

The important difference between three station machines and four station machines is the dry cycle time required.

Minimize dry cycle time

Dry cycle time comprises the opening of the molds or clamp, the index time (to accomplish 120° versus 90° of movement), and the actual close time. The open and close time should be as close to 1-5 seconds as possible. The four station machine therefore has the advantage of movement over a shorter distance.

5.2 Stretch Blow Molding

With the commercial introduction of the soft drink bottle, stretch blow molding became a commonly known term in the blow molding industry.

Biaxial stretch blow molding applies to the method of producing a plastic container from a pre-form that is stretched first in the axial direction and then in the hoop direction when the pre-form is blown into its final shape (Fig. 5.4).

Amorphous materials such as polyethylene (PE) and terephthalate (PET), which have a wide range of themoplasticity, are easier to stretch blow than semi crystalline materials such as polypropylene. When a container is produced from biaxial stretch blow molding, the properties of the raw materials are enhanced.

Enhanced material properties by stretching

These include tensile strength, barrier properties, drop impact, and clarity. A further enhancement is that a container such as a bottle can have up to a ton load.

The stretch blow molding process is based on the crystallization behavior of the resin. The pre-form is temperature-conditioned and then rapidly stretched and cooled. To give the best results, the resin must be conditioned and then rapidly stretched and oriented just above

Steps of biaxial stretch blow process

(A)

(B)

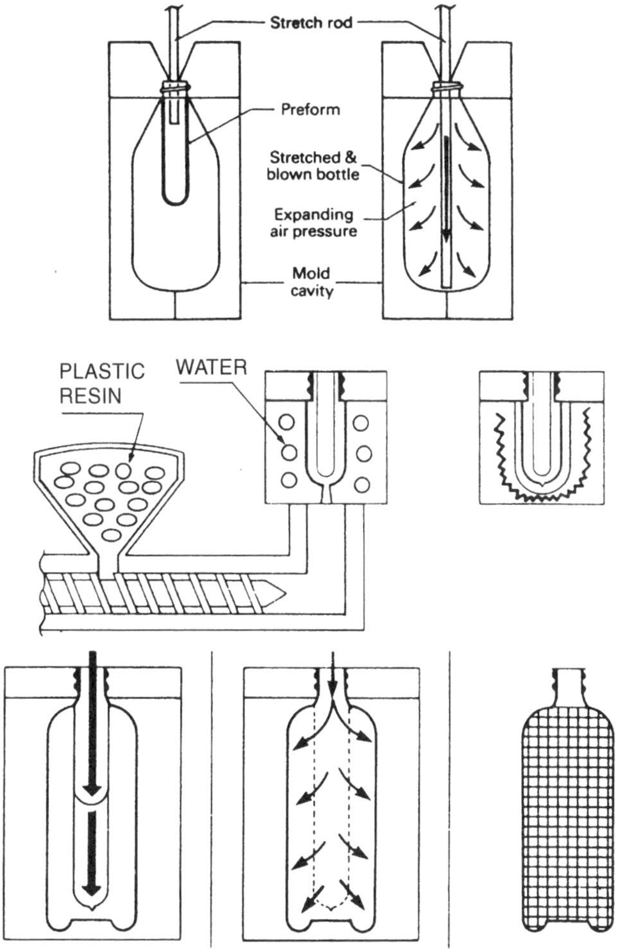

Fig. 5.4: (A) Schematic of the biaxial stretch blow process; (B) parison formation; (C) parison heating; (D) parison lengthening through use of stretch rod; (E) blow molding; (F) finished bottle

Stretching and orienting just above glass transition temperature

the glass transition temperature. It is at this point the resin may be moved without the risk of crystallization. This method is useful for molding many thermoplastic materials such as polyethylene terephthalate (PET).

As with an injection molding machine, the plastifier screw function is to convey, mix, melt, and inject (under high pressure) the homogeneous melt into the injection mold.

The old axiom, "If you make a good pre-form you will blow a good bottle," is very true. Thus the injection station is critical, which means that pre-form engineering is both art and science. Computer Aided Design (CAD) is used for the final machined design. However, hands-on experience is required for initial development.

The process uses injection molded, extruded, or extrusion blow molded parisons in one or two steps. In the one-step method, parison production, stretching, and blowing take place in the same machine, while in the two-step method the main advantage is the savings in energy.

In the one-step method, the pre-form is rapidly cooled to the stretch temperature, whereas in the two-step approach, the pre-form is cooled to room temperature. However, production of the latter method is much more efficient. Also, a minor breakdown in one of the steps does not stop the other.

Additional Reading

Belcher, S., Injection Blow Molding. In, Lee, N. C. (Ed.), *Plastic Blow Molding Handbook* Chapters 3 and 4, (1990) Van Nostrand Reinhold, New York

6 Extrusion Blow Molding Technique

Although the variety of blow molding equipment developed in recent years is impressive, the blow molding process generally consists of three stages:

Stages of blow molding

1. Melting or plasticizing the resin
2. Forming the parison or pre-form
3. Inflating the parison in a blow mold to produce the end product

6.1 The Steps of the Blowing Technique

Figure 6.1 is a schematic drawing of the steps that are the three stages of the molding cycle. The three stages in blow molding are melting the resin, forming the parison, and blowing the part, which will occur sequentially but not one immediately following the other. Trimming and secondary operations take place during the molding cycle.

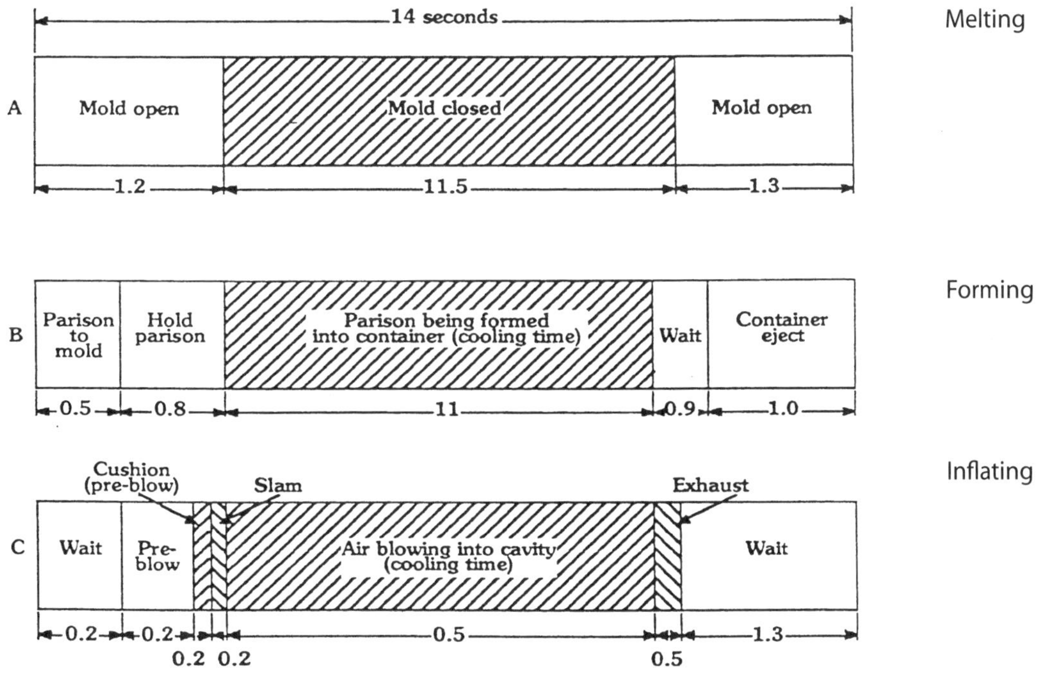

Fig. 6.1: Machine cycles

The four steps of the actual blow molding cycle take place one after another, though a number of pieces can be produced simultaneously on one machine, the four steps overlapping. Blowing takes a certain amount of time, measured in seconds or, for large items, a minute or more. Most of the molding cycle is taken up by the cooling step, and therefore, it is the cooling time which controls the blowing cycle. The speed of the machine that melts the resin and produces the parison must conform to the blowing cycle, i.e., the cooling time. Flash is removed and secondary operations such as drilling are performed during steps a, b, c, and d below.

Typical blow molding cycle

a. The molten, hollow polyethylene tube – the parison or pre-form – is placed between the two halves of the blow mold.
b. The blow mold closes around the parison.
c. The parison, still molten, is pinched off and inflated by an air blast that forces its walls against the inside contours of the cooled mold.
d. When the piece has cooled enough to become solid, the blow mold is opened and the finished piece is ejected.
e. Deflash and trim – secondary operation.

6.2 The Parison or Pre-Form

The parison or pre-form is a tube with a usually circular, but sometimes elliptic or other cross-sectional form, consisting of resin in the molten state which permits it to be blown into its final shape (stage 3). To make the parison, the polyethylene pellets or cubes have first to be melted, plasticized, and well mixed under the influence of heat and pressure, and then shaped into a tube. Figure 6.1 shows graphically a typical blow molding cycle for one-gallon containers. Three phases of the cycle are noted and each phase is broken down into its components. All blow molding cycles contain these segments, although the time intervals may vary somewhat.

As discussed previously, two techniques of melting and platicizing the resin and forming the parison are available:

Methods to produce pre-forms or parisons

1. Extrusion, which produces the parison as a kind of endless tube which has to be cut, or continuously "shot" by a hydraulic accumulator ram or from a reciprocating screw
2. Injection molding, which normally requires a mold (called a pre-form mold) to produce the pre-form, or parison, which is then transferred into the blow mold

In injection blow molding, an unsupported, suspended parison can be produced, just as in extrusion blow molding, but this is not common.

Therefore, most of the discussion in this book has been concerned with two techniques, with special emphasis on the most commonly used of the two – extrusion blow molding.

6.3 Essential Blow Molding Factors

Figure 6.2 lists the most essential factors in blow molding, including processing and operating characteristics as well as end product properties.

Most of these processing and operating characteristics and end product properties will be described and dealt with at length later on in the proper context. The aim in blow molding, as in any other industry, is to make the best possible end product with all the desired properties in the shortest possible time, under the most favorable and least expensive processing and machine conditions – that is, to produce acceptable products at the lowest possible cost.

Desirable end-product properties may be quite different, depending on the particular case. Many times, such properties as resistance to environmental stress cracking, shrinkage, impermeability, and taste, are not important, which is the case in most toys.

On the other hand, for containers used for packaging household detergents these properties are important, and so are stiffness, deterioration due to temperature fluctuations or ultraviolet radiation, discoloration, odor, piece weight as well as wall thickness uniformity, weld-line

Operation issues

Processing and Operating Characteristics

Resin flow (mainly determined by basic resin properties)
Parison sag
Parison swell
Molding cycle time
Most favorable melt temperature
Most favorable mold temperature
Blowing pressure

End product issues

End-product properties

Stiffness or flexibility (depending on what is desired)
Resistances to environmental stress cracking
Resistance to attack and penetration by chemicals
Appearances especially gloss and surface defects
Clarity
Uniformity of wall-thickness
Distortion (shrinkage and warpage)
Parting-line thickness
Weld-line shape and thickness or strength
Piece weight
Discoloration
Deterioration due to temperature fluctuations or ultra-violet radiation
Odor or taste
Toxicity

Fig. 6.2: Essential blow molding factors

strength, and parting-line thickness. But no matter what kind of hollow piece is being blown, it should always be produced at the lowest possible price, limited only by quality requirements.

6.4 Choice of the Most Suitable Resin

Resin selection

Because each end product calls for particular properties, the selection of the most suitable resin for each application is very important. Resin selection is not an easy job. The problems must be clearly defined beforehand. Careful investigation of the end-use property requirements must be made. Moreover, resin requirements vary with the blow molding procedure and the available equipment.

Quite generally, items requiring stiffness, high strength, and hardness, such as containers for household detergents or chemicals, should be made of high-density resins. (These items currently represent the bulk of the blow molding market.) For squeeze bottles and flexible items such as toys, lower density polyethylene resins are more suitable.

Taking processing into consideration

The inherent physical properties of a resin may be modified by altering processing and operating conditions during the fabricating process, starting in the extrusion or injection machine and ending in the blow mold. The main influences are those exerted by heat (or cooling) and pressure, and the time duration of these influences is essential.

That is why good knowledge of the equipment and the mold and their operation will permit their skillful use to obtain the highest possible output of high quality, uniform items. These topics are discussed in the following chapters.

Additional Reading

Rosato D. (Ed.), *Blow Molding Handbook*, 2nd ed. (2003) Hanser Publisher, Munich

Quantum U.S.I. Division. Polyolefin Blow Molding (Equistar Chemicals) and Operation Manual. Company literature

7 Guidelines for Selecting Industrial Blow Molding Machines

Today, the design of industrial blow molding equipment has become more scientific. The process is more reliable and predictable as a result of many technological advances in the equipment (Fig. 7.1).

However, there are several basic guidelines which must be considered to optimize the functioning of the equipment. As we well know there are seven major subcomponents:

Optimizing equipment

1. Extruder
2. Accumulator head
3. Press
4. Hydraulic unit
5. Pneumatic system
6. Microprocessor
7. Take-out system

We must examine each of these components to match the capacities of one to the other.

Prior to determining component sizes, it is desirable to project the fastest cycle time achievable to satisfactorily produce the part. At this point, we must also predict the total weight of the shot (finished part and trim).

Fig. 7.1: Blow molding equipment

Example

We will develop the throughput vs. process parameters for a simple case as an example.

- 3600 seconds per hour divided cycle time in seconds = yield
- Assume a 60 second cycle
- 3600/60 = 60 parts/hour the yield
- Assume a 2270 g (5 lb) shot size
- Yield × (H) number of heads × mass (shot size) = output
- 60 × 2 × 2.270 kg = 272.4 kg/h 60 × 2 × 5 = 600 lb/h

7.1 The Extruder

The extruder is the heart of the melt preparation system and must deliver a homogeneous and stable melt to the accumulator head. Typical extruder sizes for blow molding range from 64 mm (2.5 in), 76 mm (3 in), 89 mm (3.5 in), 114 mm (4.5 in), to 152 mm (6 in) diameters and usually have 24:1 or 30:1 L/D ratios.

Typical extruder sizes

Table 7.1: Extruder Sizes

Extruder size (mm)	Ratio	Power kW (hp)	Output kg/h (lb/h) at 100%	at 80%	Screw speed (rpm)
64 mm (2.5 in)	24:1	37.285 (50)	136.078 (300)	108.862 (240)	125
76 mm (3 in)	24:1	55.927 (75)	181.437 (400)	145.150 (320)	125
89 mm (3.5 in)	24:1	74.570 (100)	272.155 (600)	217.724 (480)	125
114 mm (4.5 in)	24:1	149.130 (200)	453.592 (1000)	362.874 (800)	80
152 mm (6 in)	24:1	223.700 (300)	680.389 (1500)	544.311 (1200)	70

Determine extruder size

From the example discussed earlier in this section, 272.4 kg/h (600 lb/h) are required. A 89 mm (3.5 in) 24:1 extruder has the capacity to produce at this range at 100% output. The logical choice is to size the extruder to be 114 mm (4.5 in) 24:1. This will allow for future output expansion.

During operation, 80 percent of the heat developed in the process is developed by frictional heat from the feedscrew revolving within the barrel. Therefore, it is preferable to run the extruder at lower speeds to produce the desired output.

The two most popular methods of barrel cooling are closed loop water cooling and air blowers with high air throughput. Usually for 114 mm (4.5 in) and 152 mm (6 in) extruders water cooling is desired. Seventy percent of the 89 mm (3.5 in) extruders supplied are water cooled. The 64 mm (2.5 in) and 76 mm (3 in) extruders are commonly air cooled.

In evaluation of the extruder, serious thought should be given to the gear reducer. A light-duty gearbox will not allow operation with today's high torque materials. Gear reducers have excellent service factors at high speed. However, the service factor rating is rapidly reduced at lower speeds. The ideal is to have a service factor of 1.25 at normal operating speed. The service factor is a safety range in which the gearbox can operate at a certain speed and power input. Be careful that the thrust bearing and radial bearing are sized to match the power and torque requirements of the extruder. The bearings and gears used must provide for continuous, trouble free, and long operation.

Service factor = safety range

For engineering material application, high torque at low screw rpm is required. This usually requires upgrading the gear reducer provided to one extruder size larger, i.e., on a 89 mm (3.5 in) extruder use a gear reducer sized for a 114 mm (4.5 in) extruder.

7.2 The Accumulator Head

The accumulator head design should be a first-in, first-out type. The head must be streamlined and free of internal hang-up areas. Ideally, the head will be designed with no weld or reknit problem areas (Fig. 7.2).

In selecting the shot capacity of a head, care should be taken to not oversize the head, for example, running parts weighing 1.8 kg (4 lb) on a 9.07 kg (20 lb) head. Stagnant flow areas will be developed which create dead spots within the head and will most likely produce black specks (burnt material) on the finished parts resulting in rejects. Cleaning out of the heads will take longer by using less inventory of the head on a continued basis.

Match shot size and head size

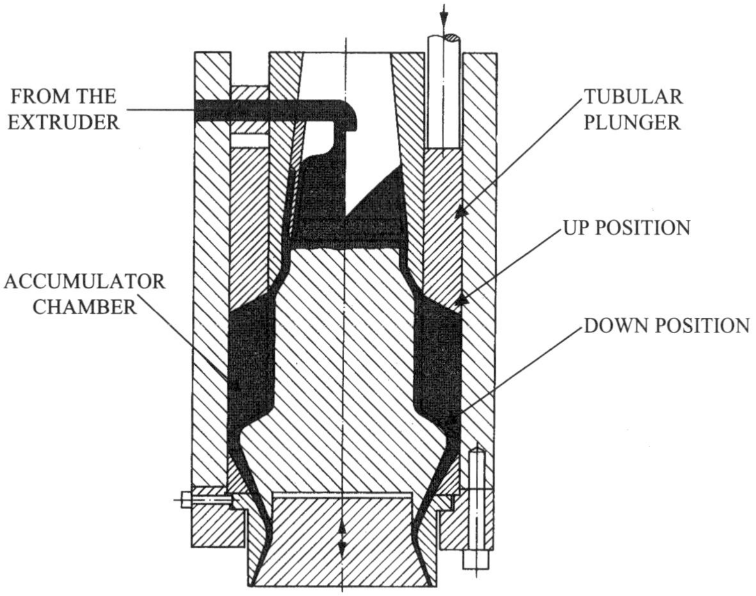

Fig. 7.2: Accumulator head (Design: Kautex-Werke, Hangelar)

Optimal tooling size

The tooling size of the accumulator head is generally restricted by the outer diameter of the main body. As an example, if the body has a 381 mm OD (15 in) for a 6.8 kg (15 lb) head, then the maximum usable tooling size is 304.8 mm (12 in).

If larger tooling is used, the program cylinder may not have the power to control the proper movement. Any larger tooling used may be restricted by the capacity of the program cylinder and the shooting cylinders. With diverging tooling, the programming area is larger, requiring more power to efficiently control the programming.

Based upon HDPE, normally, the parison swell will be greatest with small converging tooling, e.g., 16 mm (3 in) diameter. The swell at this size is around 40 percent. With 304.8 mm (12 in) converging tooling, the swell ratio may only be 20 percent. If using engineering grade material, the swell ratio may only be from 10 to 12 percent throughout the tooling size range.

Accomodate tooling change

Care should be taken regarding the method for changing tooling. If tooling is seldom changed then there is no problem. If tooling is changed often then consider quick-change tooling. There are several designs available with split two-piece being the most common and perhaps the most economical. This will allow for changing only the bottom half of the tooling, which is quicker and less expensive.

Equipment now has the ability to control push-out speeds and also fill rates. Normally, the push-out speeds can be changed for a certain position and speed. This method can be difficult to set up and control and should not be used to replace parison programming. However, push-out control will allow for fine-tuning of product wall thickness.

Fill control

Fill control of an accumulator head will provide consistent packing of the head inventory from shot to shot. This fill control should help minimize variations in parison length over a time period.

Dual head operation

Over 70 percent of the machines provided today have dual head operation. Each head must have total independence from the other. Fill rates must be controlled consistently from cycle to cycle or reject rates will be high. If using dual 16 kg (15 lb) heads, it may be required to have a shot size of 5.443 kg (12 lb) in one head and 4.082 kg (9 lb) in the other. The equipment must be designed to be able to balance the fill rates for both heads to finish filling at nearly the same time. Hydraulic proportional valves to control the fill rate of each head can accomplish this. The filling rate must be consistent from cycle to cycle.

7.3 The Press

When considering clear area, determine the actual size of the outside dimensions of the mold. However, there may also be extensions such as cylinders, water lines, clamping devices, etc., that will increase the size of the mold. Leave enough room to fit these items on the platens. If there are two molds, then there must be enough space between both molds.

Daylight openings

Pay attention to minimum and maximum daylight openings. It is extremely important that the daylight minimum and maximum openings be readily and simply adjusted to compensate for various mold sizes and shut heights. Backup plates should be considered especially with dual molds for ease of mounting and maintaining exact head center distance. Note that these plates will affect the daylight openings. Platen closing must be synchronous and parallel to produce acceptable parts. If this is not the case, there will be excessive wear on the bushings and unacceptable parts will result. Blow pin damage will be a common result of uneven closings. The overall closing of the platens should be smooth, with no hesitation or bounce-back, especially during final closing, or a bad pinch-off will result.

Choosing clamp force

The clamping force used on large industrial blow molding presses should match the actual clear area of the clamping section. Most parts are produced with a blowing pressure of between 413.685 to 689.476 kPa (60 to 100 psi) of air to actually form a part. The clamp force for any part can easily be calculated by considering the projected cross-sectional area of the part at its parting line on the mold and then multiplying this area by the blowing pressure to give the total clamp force needed. A clamp force of 27215.542 kg (30 tons) would be required on a part having a cross-sectional area of 3870.96 cm^2 (600 in^2) and a blowing pressure of 68.947 kPa (100 psi).

It is not uncommon to have a blowing pressure in excess of 103.421 kPa (150 psi). This will greatly increase the tonnage requirement as indicated above from 27215.541 to 40823.313 kg (30 to 45 tons). This must be considered prior to designing the press section. (See also Section 10.7 for the discussion of clamping needs in blow molding.)

7.4 Summary

Early communications between customer and all suppliers

Industrial blow molding systems are very complex. Care must be taken prior to placing orders for equipment to properly size all of the major subcomponents. Upon selection of the machinery supplier, a kick-off meeting with the customer, material supplier, mold builder, and machinery supplier should be held to finalize all aspects of the blow molding project. This must come early in the timetable. See the sequential series of illustrations showing blow molding press considerations in Fig. 7.3.

(A)

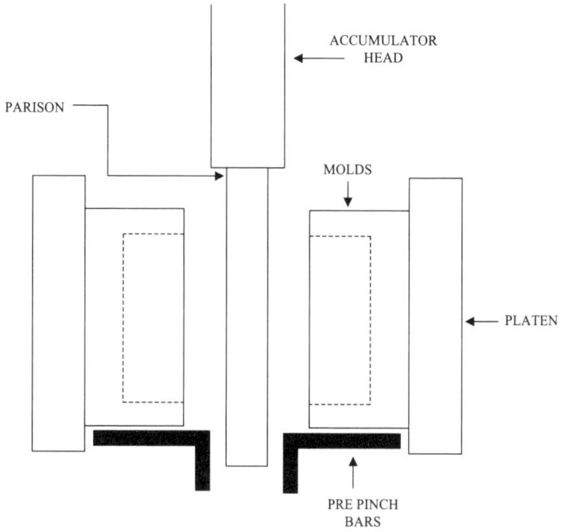

Fig. 7.3: Sequential diagrams of the blow mold process

(B)

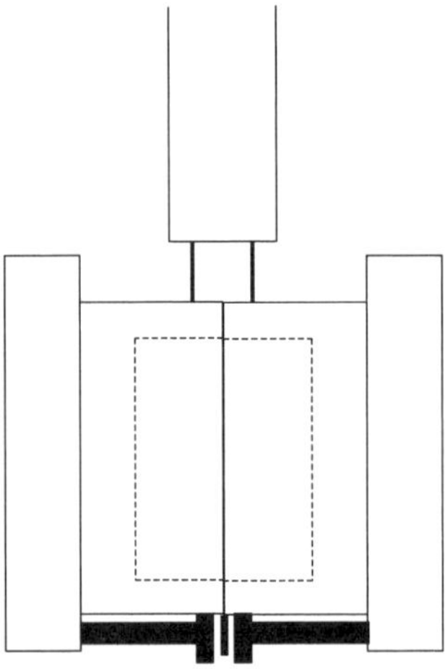

(C)

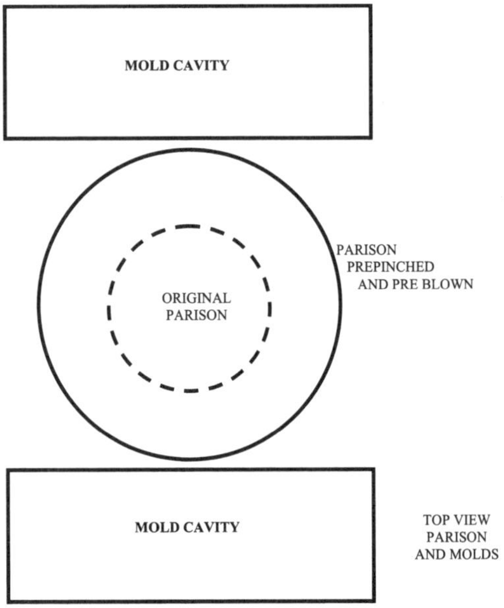

Fig. 7.3: Sequential diagrams of the blow mold process (continued)

(D)

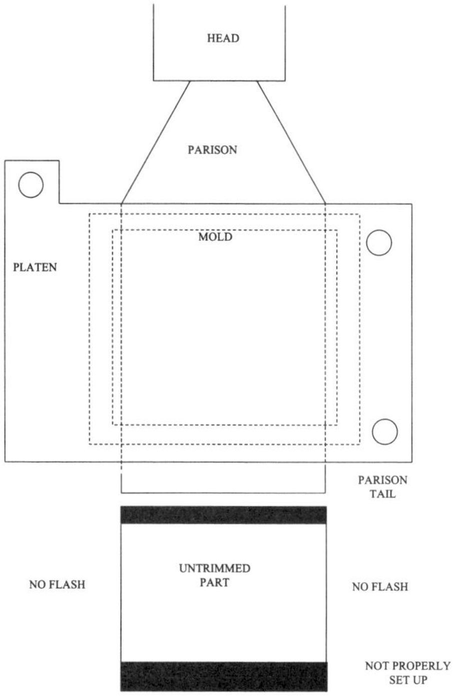

(E)

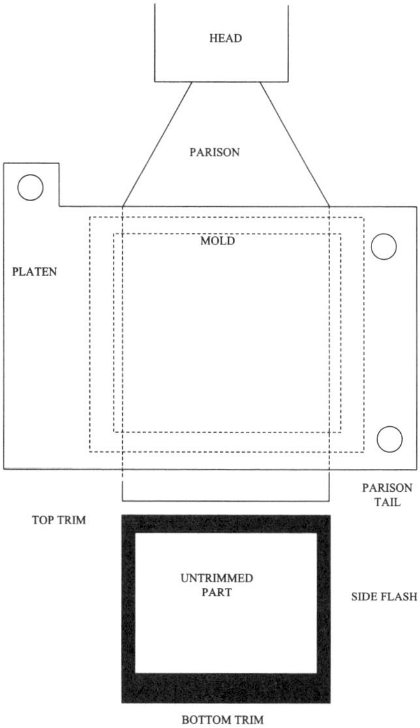

Fig. 7.3: Sequential diagrams of the blow mold process (continued)

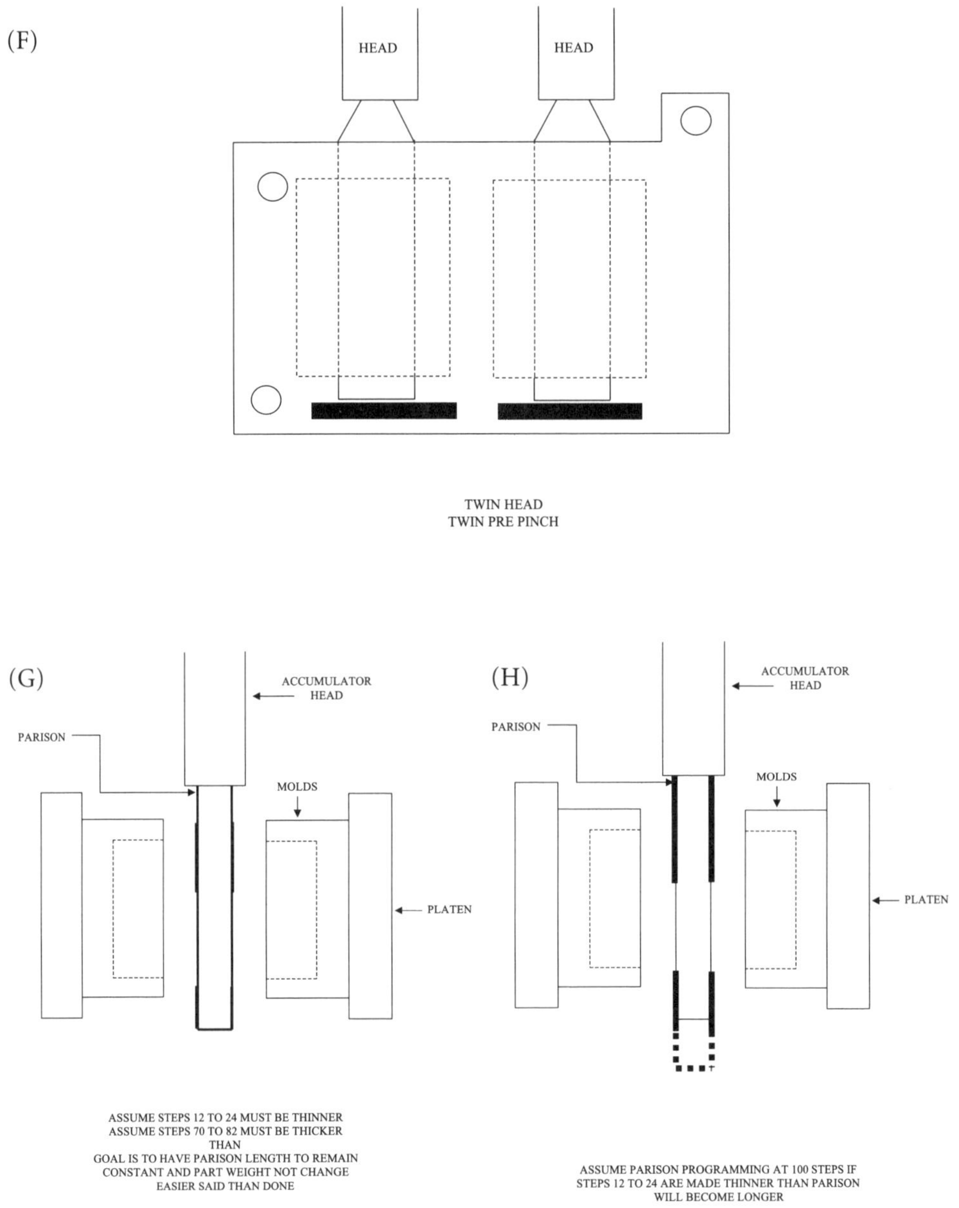

Fig. 7.3: Sequential diagrams of the blow mold process (continued)

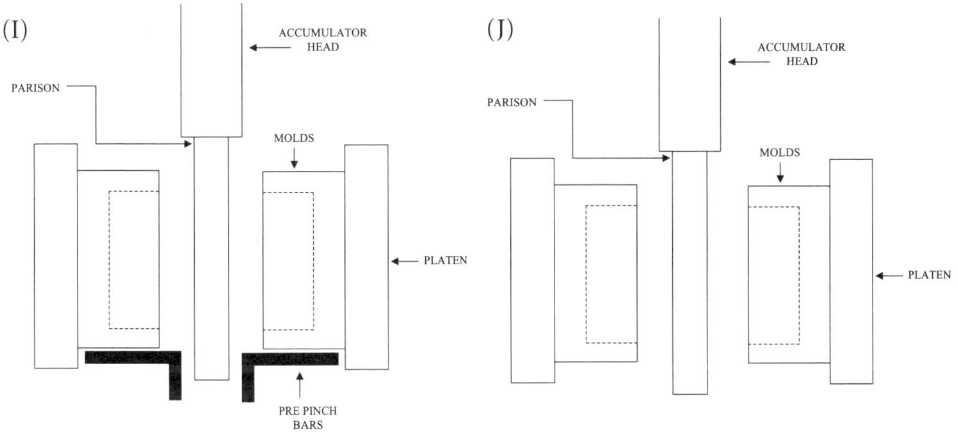

Fig. 7.3: Sequential diagrams of the blow mold process (continued)

Additional Reading

Hsu, John, SPE *Blow Molding Seminars Presentations and Manuals*. Brookfield, Connecticut (1993-1999)

Slawska, R., SPE *Blow Molding Seminar Presentations and Manuals*. Brookfield, Connecticut (1993-1999)

8 Extrusion Blow Molding Advanced Systems

8.1 Coextrusion Blow Molding

Barrier properties by multi-layer blow molding

Coextrusion of several materials has been around for some time and entered the blow molding industry more than 20 years ago. With its increasing success in converting from glass and metal to plastics, the industry continues its search for plastic materials and processes to introduce more plastic packaging applications. Multi-layer blow molding products seem to penetrate the higher-volume glass and metal container market for those applications that can be switched to a plastic container. The billions of containers used each year for food and chemical packaging require barrier properties that cannot be accomplished with most plastic resins. The food industry needs either an oxygen barrier or a moisture barrier in their containers so that their products can be placed on the shelves without refrigeration. In addition, the chemical industry needs chemical barriers for packaging their products in plastic containers. All these requirements can only be met by using multi-layer combinations of different plastic resins to produce plastic containers. Figure 8.1 shows an example of a multi-layered blow molded bottle.

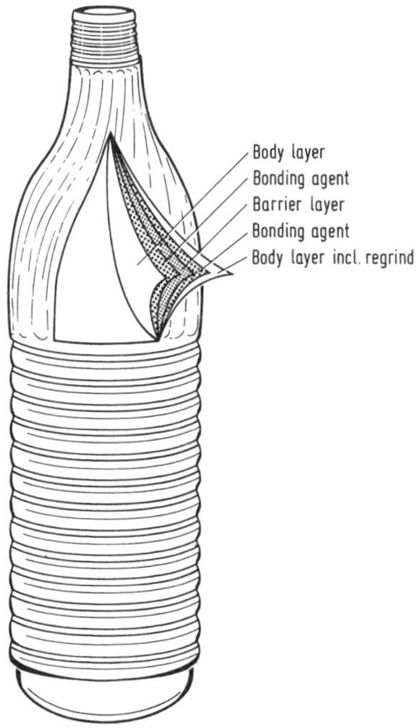

Fig. 8.1: Five-layer extrusion blow molded bottle

Fig. 8.2: Large coextruded parts (Courtesy of Kautex Machines, Inc. North Branch, N.J.)

Machines for coextrusion

Coextrusion is an extrusion blow molding process utilizing several extruders instead of one. Proprietary wheel type machines as well as platen type machines are used in industrial production. The advantage of coextrusion is that equipment and technology are both readily available; it is fast and economic.

8.1.1 Common Materials and Applications for Coextrusion Blow Molding

Materials for coextrusion

Polypropylene (PP) is predominantly used as the base material, mainly because most food products that are packaged in coextruded bottles need to be hot-filled and PP can withstand the hot-fill temperatures. PP provides contact clarity but cannot be manufactured into a crystal clear container. Polycarbonate combined with EVOH, for example, can be used to produce a container that is crystal clear and has an oxygen barrier. The viewing stripe bottle is another example of an coextruded bottle that has been around for many years. In order to manufacture a clear viewing stripe in a colored container, coextrusion with a mini-extruder in combination with the base extruder for the body material is used.

Acrylonitrile is an example of a plastic material in blow molding that can be used on the inside of plastic containers to provide a barrier for chemicals or pesticides.

Coextrusion blow molding is the method used to make products that have multiple layers in their wall structure. The layers may be the same resin but different colors or different resins with adhesive layers between them.

Coextrusion is used for the following applications.

Applications of coextrusion

- Multi-colors
- Decoration
- Viewing stripe in containers
- Barrier properties provided by several resins with differing characteristics
- Recycling reclamation by use of 2 or 3 layered structures

Figures 8.2 and 8.3 show examples of applications made by the coextrusion method.

Typical coextruded parts

Fig. 8.3: Large coextruded parts (Courtesy Kautex Machines, Inc. North Branch, N.J.)

8.1.2 Typical Structures

Two, three, and six layered structures and even seven layers are in production for large part coextrusion. Environmental, economic and recycling concerns are accelerating the need for improved barrier properties (e.g., in plastic fuel tanks) and will represent the largest growth sector in the future. Figures 8.4 and 8.5 illustrate a variety of structures and applications.

Design of multi-layered parts

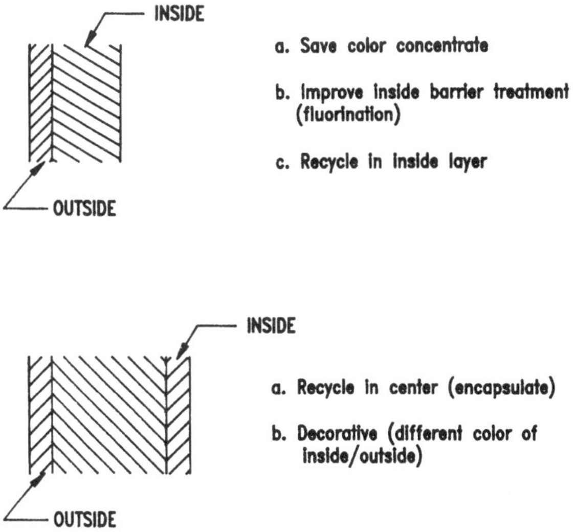

Fig. 8.4: Two layer/three layer cross- section (Courtesy Kautex Machines, Inc. North Branch, N.J.)

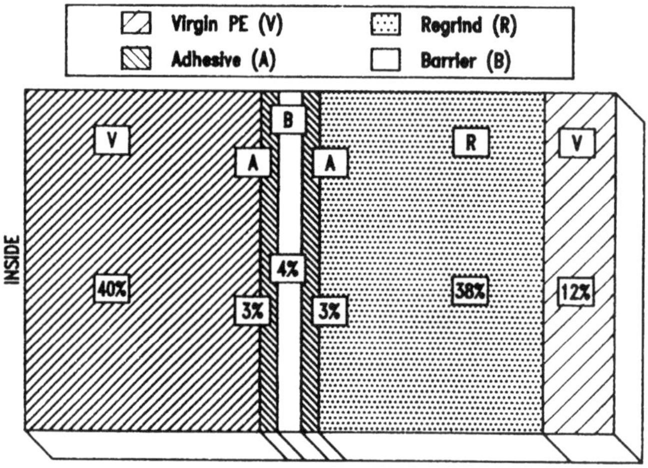

Fig. 8.5: Typical fuel tank layer thickness in barrier coextrusion. (Courtesy Kautex Machines, Inc. North Branch, N.J.)

8.1.3 Continuous Multi-Layered Coextrusion

Enhancing flow and simplifying machine design and control

Typically, the continuous extrusion method is used because it enhances the laminar flow of the various components that make up the composition of the die head. It also greatly simplifies the machine design and controls.

A schematic layout of a six layered machine, with reciprocating extruders that deliver the resins to a multilayered head is shown in Fig. 8.6.

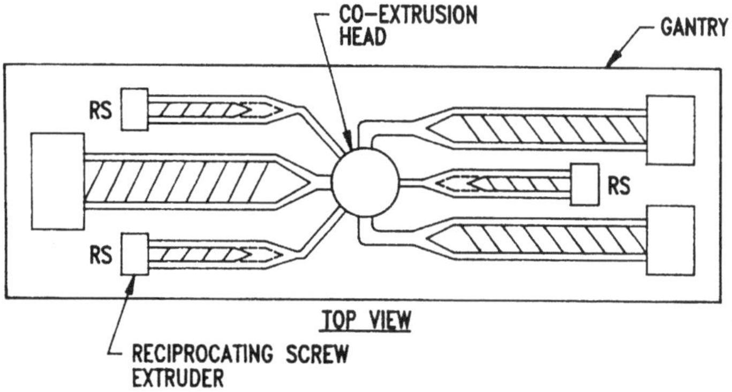

Fig. 8.6: Location of extruders to coextrusion head (Courtesy Kautex Machines, Inc. North Branch, N.J.)

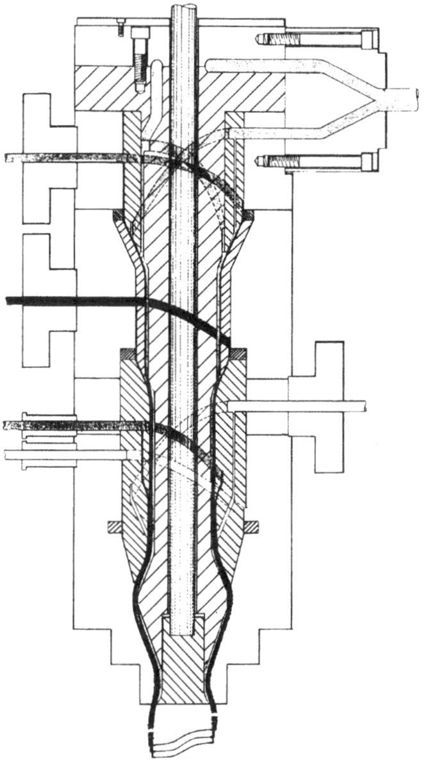

Fig. 8.7: Kautex six-layer coextrusion die with cadoidal manifolds (Courtesy Kautex Machines, Inc. North Branch N.J.)

The design for a six layered head is shown in Fig. 8.7

The extrusion system in coextrusion blow molding must supply several streams of melted resin to the die simultaneously. Some streams are smaller, by design, than others in order to produce thinner layers.

8.2 Three-Dimensional Blow Molding

Minimizing flash and improving quality

Three-dimensional blow molding was developed several years ago in Japan and refined further in Germany in recent years. Parts with irregular shapes manufactured with conventional blow molding methods create large areas of flash, whereas a seamless molded part has flash only at the top and bottom of the part (see the example of an air duct in Figure 8.8).

The advantages of the reduced flash are that it allows the use of a smaller extruder, uses less power, and smaller grinding equipment and thus less regrind handling. Other advantages include: even wall thicknesses, improved mechanical strength, no finishing at the outer edges, resulting in higher product quality and lower clamp forces.

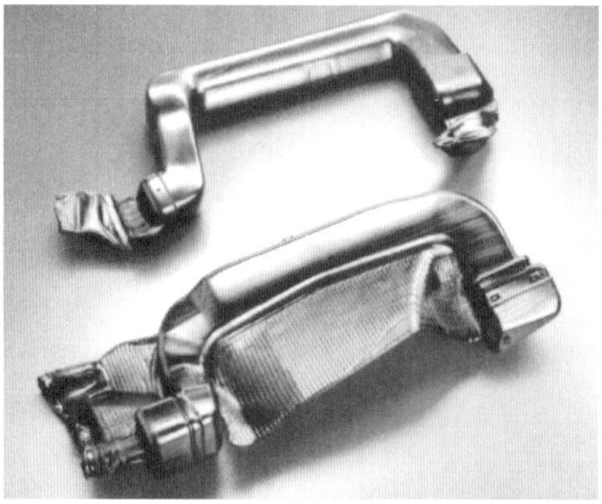

Fig. 8.8: Air duct parts. Top: three-dimensional blow molded part, bottom: conventional blow molded part (Courtesy of Kautex Machines, Inc. North Branch, N.J.)

8.2.1 Three-Dimensional Blow Molding Processes

There are three types of three-dimensional blow molding technology available:

Various types of processes

1. The suction blow mold module,
2. Vertical clamp parison manipulation, and
3. Horizontal segmented mold parison manipulation.

8.2.1.1 Suction Blow Molding

Figure 8.9 shows the four phases of the suction blow molding process.

Using this process, the parison is sucked into the closed mold, requiring ejection of the parison and extraction of the air volume.

The blow mold consists of a main section as well as an upper and a lower sliding core segment. The process consists of the following steps:

Process steps

1. The mold closes.
2. The suction device, which is part of the bottom segment, is ready once the mold is closed.
3. The parison ejection process begins simultaneously with the suction function.
4. The parison is preblown with support air.
5. As the ejected parison reaches the desired length, the ejection and suction processes are stopped automatically.
6. The article is blown by a blow pin or needle.

At the end of cooling time the mold opens and the article is removed.

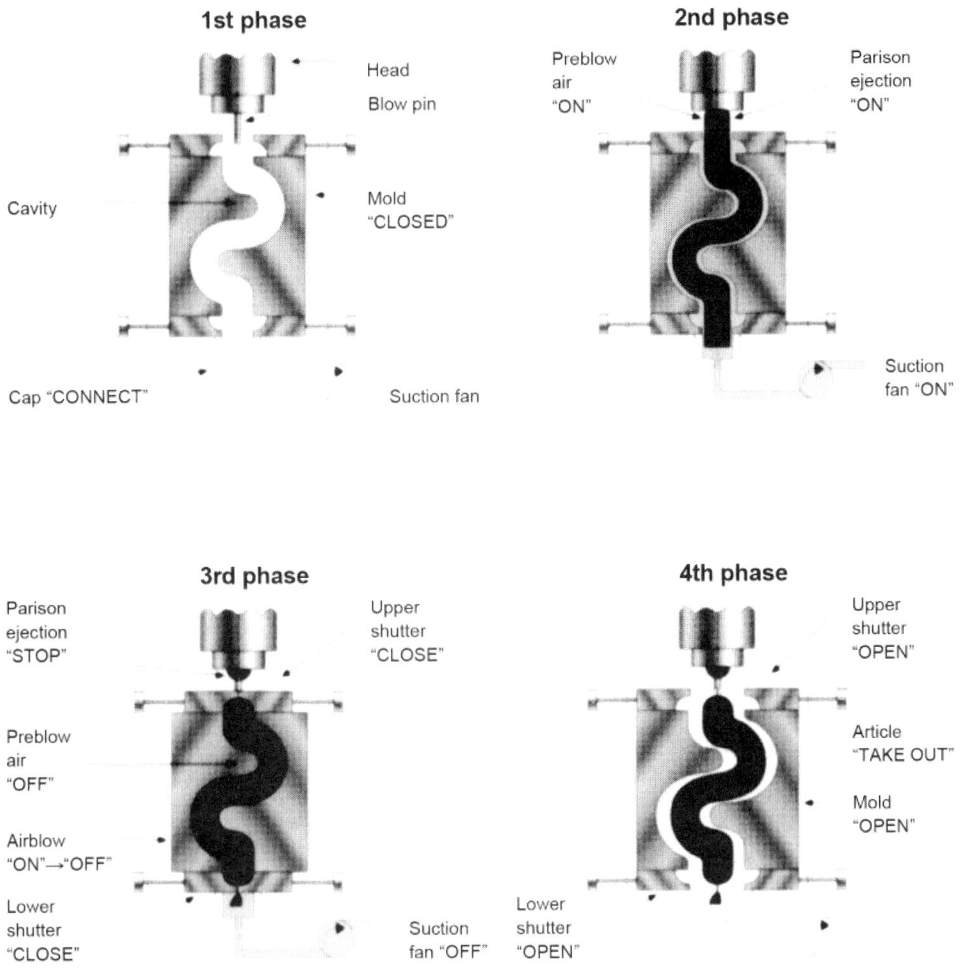

Fig. 8.9: Suction blow molding (Courtesy Kautex Machines, Inc. North Branch N.J.)

8.2.1.2 Parison Manipulation

In the vertical clamp method, the mold opens vertically, the lower half slides out, the parison is placed in the cavity, and the mold slides back and blows. This method is ideal for multi-layered applications, such as fuel filler pipes. Conventional mold halves provide for a lower mold cost. The mold is very accessible, and double clamp stations can be used for higher volumes (see Fig. 8.10).

In the segmented mold/horizontal clamp process (see Fig. 8.11), the parison is extruded and as the mold begins to close, the parison is inflated. Then, the parison is manipulated to the cavity configuration and the mold completes closure, cools, and opens with part ejection. The parison manipulation is accomplished with a six-axis robot. Complex shapes can be produced by using a segmented mold technology, with the incorporation of value-added design features through the use of a clamp function and pinch-off twin tubes, brackets, and so on.

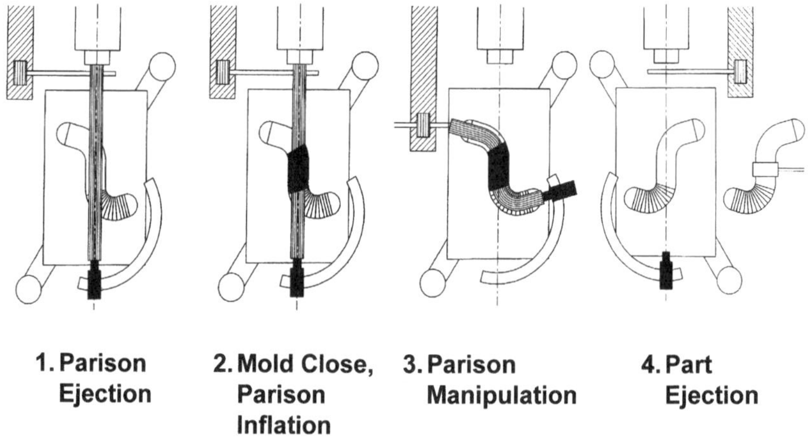

Fig. 8.10: Integrated blow molding (Courtesy Kautex Machines, Inc. North branch, N.J.)

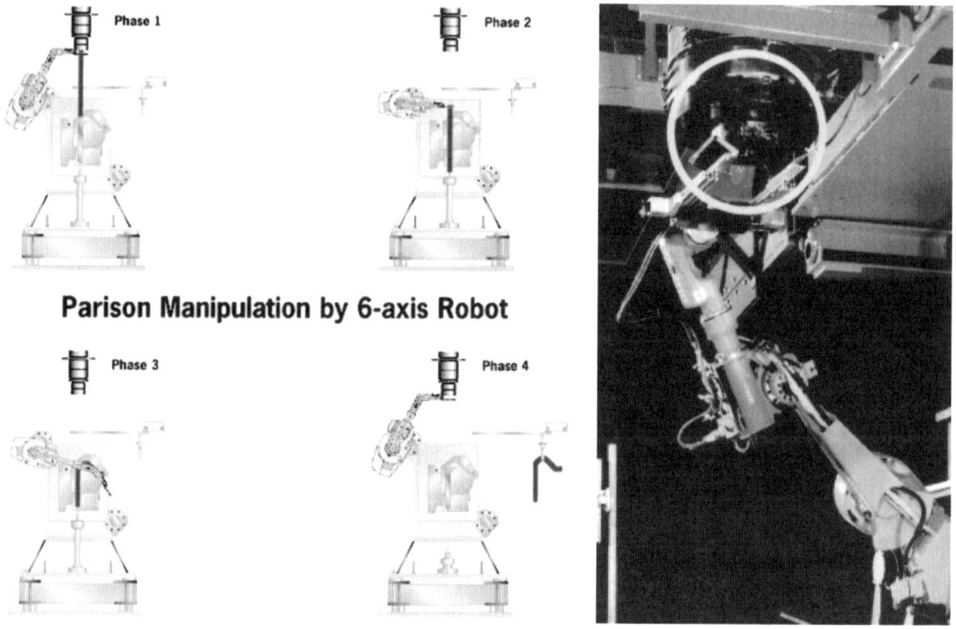

Fig. 8.11: Horizontal clamp (Courtesy Kautex Machines, Inc. North Branch N.J.)

9 Understanding Extrusion Blow Molds

Determining shape and appearance

The mold determines the shape of the end product with all its details. It helps provide the end product with essential physical properties and the desired appearance. Usually, the mold maker builds the blow mold according to the molder's or the customer's specifications, but frequently, minor adjustments or improvements which would not justify its return to the mold maker, can be made with equipment and knowledge available in the blow molding shop.

9.1 Main Characteristics of the Mold Halves

Major features of blow molds

The blow mold may have a number of parts, counting its various inserts, but it usually consists of two halves. When closed, these halves will form one or more cavities which will enclose one or more parisons for blowing. The two mold halves are usually alike. There are usually no male and female sections.

Pinch-off edges are generally provided at both ends of the mold halves. A blowing pin may have the additional function of shaping and finishing the neck inside.

Both mold halves must have built-in channels for the cooling water. Sets of guide pins and bushings or side plates in both mold halves ensure perfect cavity alignment and mold closing. Accurate guiding devices in both mold halves reduce setup time. Figure 9.1 shows the two halves of a blow mold for small bottles. Figure 9.2 shows the location of the cooling water channels.

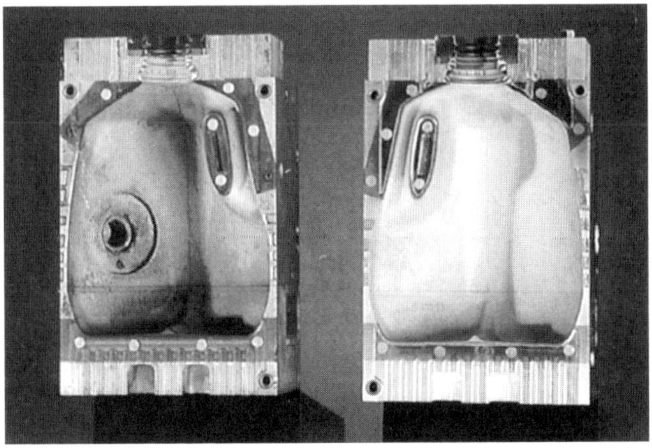

Fig. 9.1: Milk bottle mold halves. Courtesy of Johnson Controls Inc., now Uniloy Division, Milacron, Manchester, MI

Different types of cooling channels

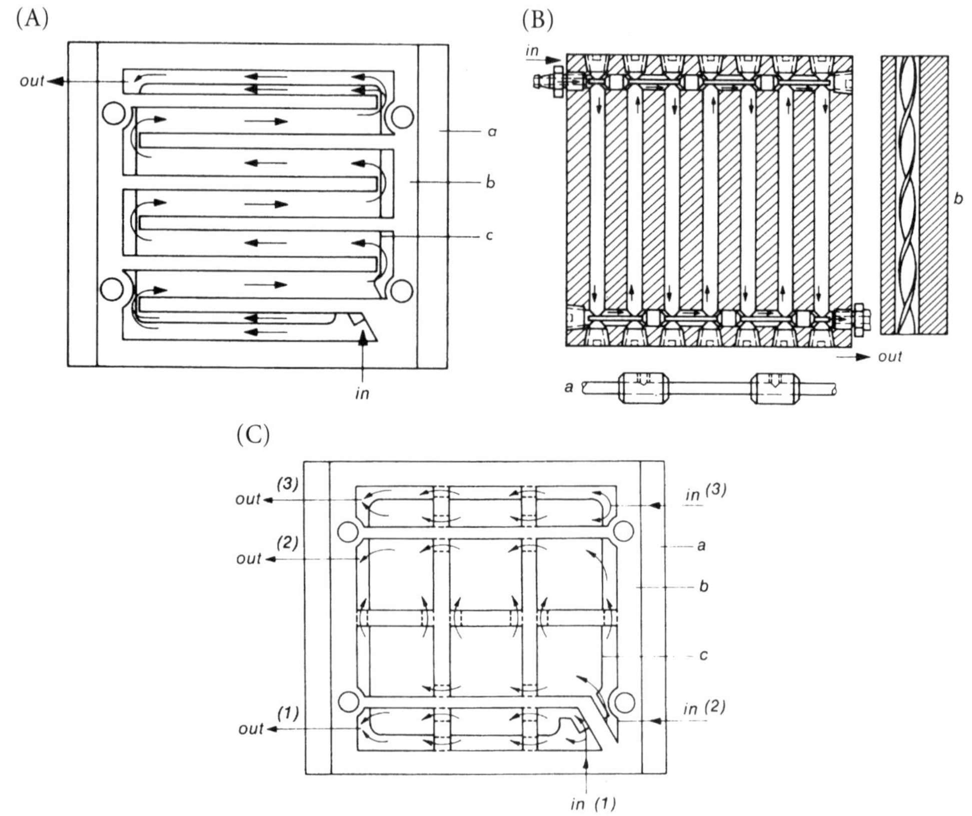

Fig. 9.2: (A) Cooling a blow mold by means of three cooling circuits and cooling chambers: a, mounting plate; b, blow mold; c, visible contours of mold cavity for a canister with a filler connection at one corner. Courtesy of Hoechst AG

(B) Cooling channels with labyrinth-type water flow: a, continuous rod with stoppers; b, inserted copper spiral. Courtesy of Hoechst AG

(C) Cooling a blow mold by means of three cooling circuits and cooling chambers: a, mounting plate; b, blow mold; c, visible contours of mold cavity for a canister with a filler connection at one corner. Courtesy of Hoechst AG

Mold closing

On some blowing presses, mold closing is carried out in two steps, first at high speed, with lower pressure, say, 6.5 to 13 mm (0.25 to 0.5 in) of daylight. The second step is slower with higher pressure to protect the mold from tools or anything else that might have fallen between the halves (which, however, should never happen in a well-kept shop) and for operator safety.

Molds are not necessarily positioned vertically, that is, in line with the parison. They may occasionally be tilted (Fig. 9.3). This will result in a nonuniform distribution of resin which may be helpful, for instance when such irregular pieces as a pitcher with a handle are being blown. It may also result in some savings in parison length.

Mold for irregular parts

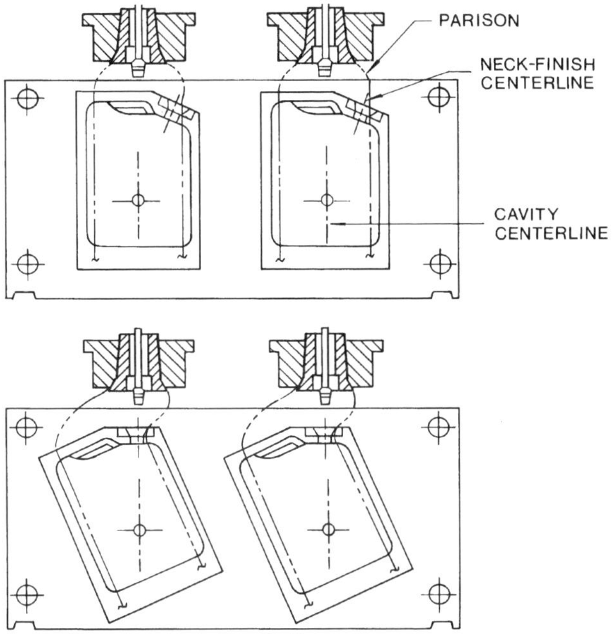

Fig. 9.3: Rotating molds (shown open at the top and closed below) are used for containers with finish centerlines at an angle to the container centerline. Courtesy of Johnson Controls (now Uniloy Division Milacron) Manchester, MI

9.2 Mold Materials

Typical mold materials and their applications

Because of the comparatively low clamping and blow pressures, the blowing mold need not be made of a high tensile strength material, with the possible exception of molds for very long production runs, say, hundreds of thousands or millions, which are sometimes made of steel. The predominant raw materials for blow molds are machined from aluminum billet, cast aluminum alloys, zinc alloys such as kirksite, and occasionally, bronze.

Beryllium copper, because of its expense and difficulty in machining, is usually reserved for pinch inserts or cores where rapid heat transfer is needed. All these alloys are excellent materials for blow molds. Table 9.1 compares the properties of common mold material.

Aluminum is the softest of the mold materials in use and is most easily damaged in the shop. Furthermore, aluminum molds wear easily. On the other hand, they are easiest to machine. Aluminum and beryllium copper cast molds may be slightly porous, and occasionally, blow molders have experienced some permeability of such molds to the viscous resin. This may affect the appearance of the blown part. The remedy for this problem is coating the inside of the mold halves with a sealer (such as radiator sealer). This will not affect the heat transfer between the resin blown against the mold and the mold walls.

Steel molds are heavier, more expensive, and more difficult to machine than those made of nonferrous alloys. Higher weight will mean more setup time in the molding shop. Moreover, the heat conductivity of steel is inferior to that of the three nonferrous mold materials. This results in a slower cooling rate and a correspondingly longer cooling cycle and consequently, a lower production rate for steel molds.

Table 9.1: Properties of Common Mold Materials

	Thermal conductivity (watts/(cm²) (°C/cm)) (BTU in/ft² H°F)	Density kg/ cm² (lb/in³)	Hardness[a]	Tensile Strength kPa (psi)	Composition
Beryllium copper alloy					
CA 172	1.11034 (770)	82.4864 (0.298)	37 RC	1.0342×10^6 (150.000)	1.8% Be
CA 824	1.08150 (750)	84.1472 (0.304)	34 RC	9.3079×10^5 (135.000)	1.65% Be
Aluminum alloy					1.6% Cu, 0.23% Cr
7075 T6	1.29780 (900)	27.9568 (0.101)	150 BR	4.9642×10^5 (72.000)	5.6% Zn, 2.5% Mg
AISI P = 20 steel			30-60 RC		0.35% C, 1.0% Ni
	0.28840 (200)	78.0576 (0.282)	290-330 BR	8.2737×10^5 (120.000)	and 1.0% Cr
AISI 420 stainless steel					0.38% C, 0.8% Si
	0.23937 (166)	77.5040 (0.28)	50 RC	1.4686×10^6 (213.000)	13.6% Cr, 0.5% Mn, 0.3% V
Kirksite A	0.92288 (640)	69.20 (0.25)	100 BR	2.4132×10^5 (35.000)	contains Cu, Al, Mg

[a] RC, Rockwell C; BR, Brinell

9.3 Importance of Fast Mold Cooling

Cooling time determines cycle time

Fast heat transfer of the material of which the mold is made is of utmost importance because, as has been explained, the cooling step controls the length of the blow molding cycle. (Cooling takes up roughly two-thirds of the entire blowing cycle.) Good heat transfer means faster cooling, and faster cooling means more items blown per hour, that is, less expensive production. This is the main reason why, for blowing molds, the above mentioned alloys are generally preferred to the usually more durable steel.

Considering only their heat transfer rate, the principal blowing mold materials follow each other in this order:

- Beryllium copper
- Aluminum
- Kirksite
- Steel

Best heat transfer rates

Occasionally, several different alloys are used in the same mold to obtain desired strength and special cooling conditions. However, as these mold materials have different heat transfer rates, a blow mold, with the exception of steel pinch-off inserts (see Section 9.4), should be made of only one material. Different materials with consequently different heat conductivity at various points of the mold will result in nonuniform cooling. This, in turn, might set up areas of stress in the finished piece, which would be susceptible to splitting in use. The blow mold halves must always be adequately cooled to solidify the part quickly, immediately after the parison has been blown out against the mold walls.

Water cooling techniques

The cooling water may be tap water. If it has a high content of minerals which may settle in the narrow cooling channels, a closed system for circulating purified water should be used. Unless the water is cold enough, as for instance, in winter, it should be chilled by a heat exchanger to 4 to 20 °C (40 to 70 °F). Such low temperatures may, however cause water condensation on the outside mold walls. Some molders, though, use noncooled tap water. Usually, the cooling water is recirculated, that is, reused time and again for a long period. Sometimes it is partly recirculated and mixed with fresh tap water to maintain the desired temperature and to economize.

The water usually circulates through the hollow mold halves. Sometimes, a copper tubing system is cast into the mold. However, to create the most useful flow, water channels are machined into the material cast into the mold halves. Well-placed channels will ensure that the cooling water comes as close to the mold cavity as feasible (see Fig. 9.2). Cooling channels should also be as close to the (lengthwise or other) parting lines caused by the separation lines of the two halves or by inserts. Parting lines will practically always show along mold separation lines. Cooling these areas will result in better finish of the piece along the parting lines.

Multiple cooling zones

Larger molds may be equipped with several – up to three or more – independent cooling zones. Generally, in the top or bottom areas, that is, around a bottleneck or the bottom pinch-off, or both, greater masses of resin are required than along the other areas. Such areas as well as thicker wall sections, therefore, often require additional cooling. Otherwise, these sections would still be viscous while the thinner wall sections have solidified when the piece is ejected. This may result in a deformed piece or one with nonuniform shrinkage and resultant warpage, which the customer will reject. That is why frequently even a simple mold has two or more cooling systems for each half.

Cooling the blowing pin

Occasionally, the blowing pin is also being cooled. Sometimes, air cooling from the outside is provided for the pinched "tail" of the parison sticking out of the mold bottom. The tail is much thicker than the wall and cools correspondingly slower.

Air may be circulated inside the blown part to speed up its cooling.

Cooling time is strongly affected by the extrusion melt temperature of the blow molding cycle. It has been experienced that an increase, or decrease, of 5 °C (10 °F) in melt temperature can extend, or shorten, the cooling cycle by as much as one second.

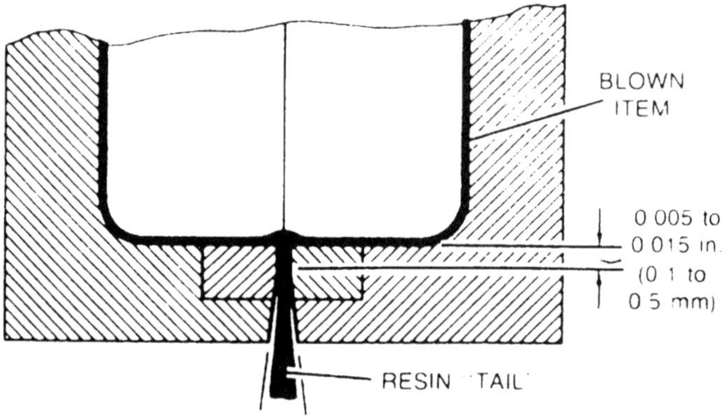

Fig. 9.4: Parison tails

9.4 The Pinch-Off

Because of the comparatively high pressure and mechanical stress exerted on the mold bottom when (in the closing step) it pinches one end of the parison together, the pinch-off in a nonferrous metal mold is frequently an insert made of hard, tough steel. The effect on the blown part always shows in the so-called weld line.

Select optimal material and design for pinch-off insert

The pinch-off section does not cut off the excess parison tail (see Fig. 9.4). Its protruding edges cut nearly through, creating an airtight closure by pinching the parison along a straight line which makes it easy later to break off or otherwise remove the excess tail piece. A high quality pinch-off of a thick-walled parison is more difficult to obtain than that of a thin-walled parison. However, much depends on the construction of the pinch-off insert.

The pinch-off should not be knife-edged, but, according to some molders, should be formed by lands about 0.1 to 0.5 mm (0.005 to 0.015 in) long. The total angle outward from the pinch-off should be acute, up to 15° (Fig. 9.4). These two features combine to create a welding line, which is rather smooth on the outside and forms a flat elevated line or a low bead inside, not a groove. A groove, which weakens the bottom along the seam, may be formed when these two features of the pinch-off are missing. Figure 9.5 shows a cross section through a well-shaped container bottom and one through a poor one due to an incorrectly constructed pinch-off insert.

One method of obtaining more uniform weld lines is to build "dams" into the mold halves at the parison pinch-off areas. These dams force some of the molten resin back into the mold cavities to produce strong, even weld lines (Fig. 9.6).

Fig. 9.5: (A) A good pinch-off result; (B) a groove results from a bad pinch-off

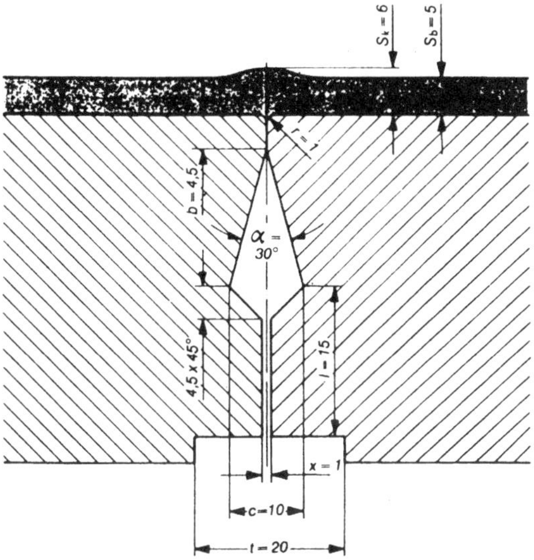

Fig. 9.6: Optimized dam for a 60-liter can made from HMW-HDPE (Courtesy of Hoechst AG)

9.5 High Quality Undamaged Mold Cavity Finish

Surface of end product determined by mold finish

A high quality mold cavity finish and undamaged inside surfaces are essential in polyethylene blow molding to avoid surface imperfections in the end product. If the highest possible gloss of the end product is desired, the mold cavity should be sandblasted with 100-grit flint sand, and there should be vacuum assists for the removal of entrapped air. If other endproduct finishes are desired, the mold cavity should be finished accordingly. Even a first-class machining job inside the mold cavity cannot prevent the occurrence of parting lines, especially if the blown item has a very thin wall. See Fig. 9.7 for an illustration of a complex machining job.

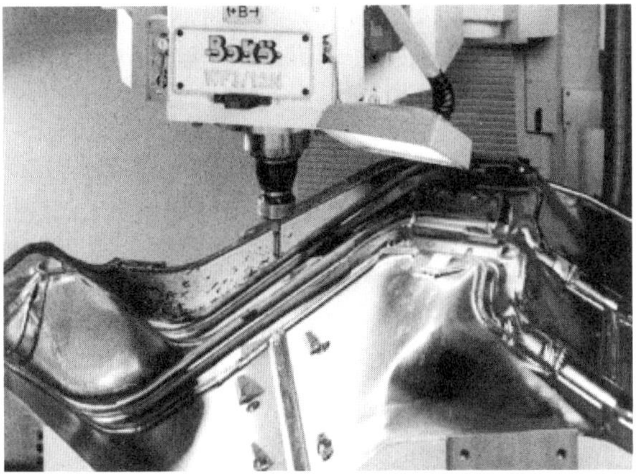

Fig. 9.7: Machining a complex part

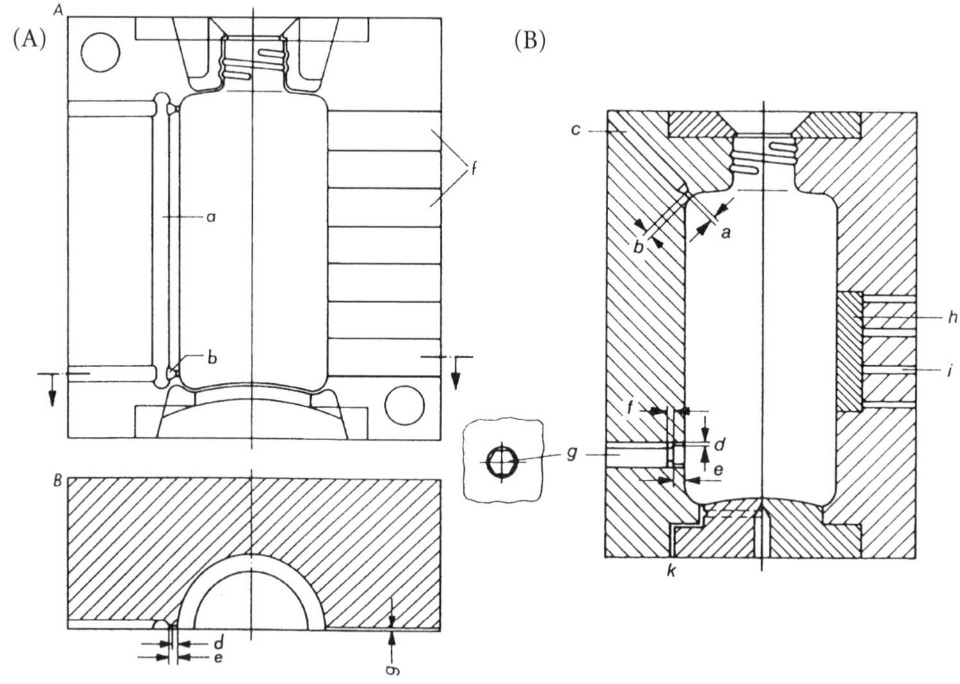

Fig. 9.8: Vents

9.6 Effects of Air and Moisture Trapped in the Mold

Proper venting for good part surfaces

In highly polished molds, air may be trapped between the mold walls and the hot, still soft piece, marring the surface of the piece. This will happen especially when thickwalled, large pieces are blown. In such cases, the mold must be vented by either sandblasting – resulting in a matte outer surface of the piece – or by grooves in the separation lines or, in extreme cases, by valves in the mold.

Generally, about one-half of the parting line periphery is vented to a depth of 0.05 to 0.102 mm (0.002 to 0.004 in). In venting difficult areas, such as handles or thread inserts, holes are usually drilled into these areas so that they vent to the atmosphere. The vent holes are normally about 0.204 to 0.254 mm (0.008 to 0.010 in) in diameter. Particular care must be taken when drilling these holes so that the mold cooling cavity is not pierced (Fig. 9.8 (A) and (B)).

Moisture in the blowing air may result in marks on the inside of the blown part. This may result in high reject rates, especially if the part is transparent or translucent. Moisture in the blowing air can be removed by means of a heat exchanger which cools the compressed air, or by traps and separators in the pipelines.

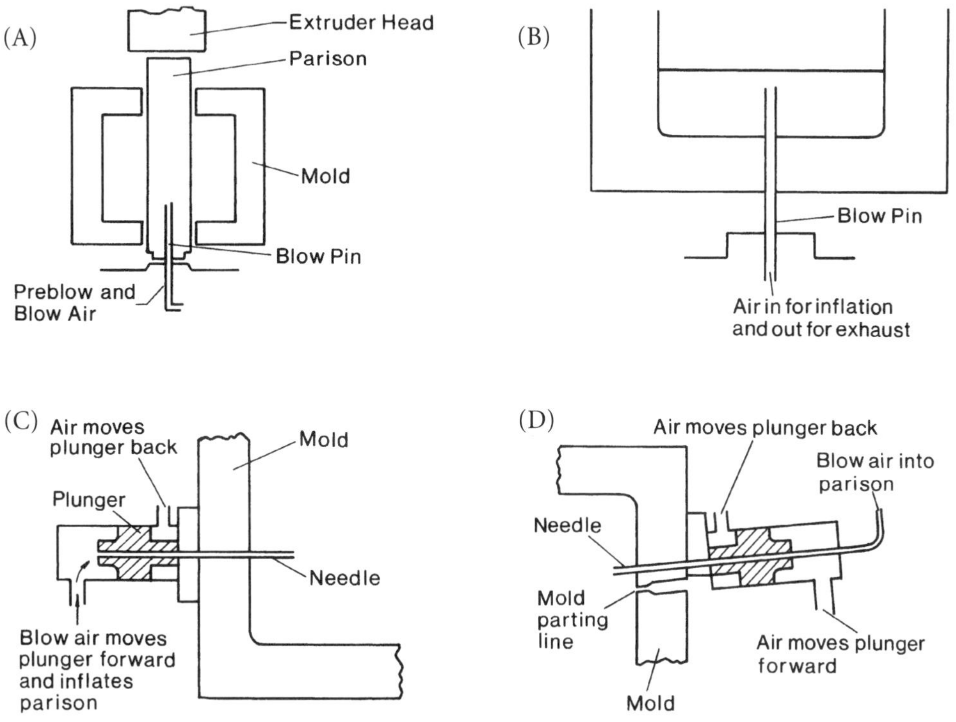

Fig. 9.9: Blowing methods

9.7 Injection of the Blowing Air

Injection of blowing air can be done by various means, such as downward through the core, through a blowing needle inserted sideways through the mold wall, or from below through a blowing pin moved up into that end of the parison which will become the (frequently threaded) neck of a bottle (Fig. 9.9 (A), (B), (C), (D)).

Different ways to inject blowing air

Sometimes, different blowing devices are used in combination. Like every step in the blow molding cycle, blowing time and duration must be well coordinated with all other parts of the cycle. As explained before, compared with the cooling time, blowing time is very short.

To obtain rapid inflation of the hollow piece, the volume of injected blowing air should be as large as possible. The opening through which the air enters the mold must, of course, be adequate.

The thinner the wall thickness and the lower the melt and mold temperature, the faster the blowing rate should be and the higher the blowing pressure, up to about 10.5 kg/cm^2 (150 lb/in^2) for very cold molds and thin-walled parts. (Compressed air injected into a cold mold may lose some pressure because cooling gas contracts.) High blowing pressure requires correspondingly high clamp pressure to keep the mold tightly closed during the blowing step.

Wall thickness and temperature influence blowing rate and pressure

9.8 Ejection of the Piece from the Mold

Automatic removal of parts

Ejection of the blown piece can be effected forward between the mold halves or downward, provided the press is built in such a way that a free fall from between the open mold halves is possible. Many machines have an automatic ejector, or stripper, assembly. Ejector, or knock-out, pins or plates push the piece out with a rapid motion, preferably hitting it on a trim area so that the piece itself will not be distorted. Ejection of the blown piece is part of the automatic blowing cycle.

If the piece is not too large, it can be blown forward out of the mold by an air jet from behind.

Manual removal of parts

The operator sometimes manually removes very large pieces from the mold. To reach the piece, the operator must first push the protective gate or shield aside, which automatically stops every mold or platen movement so that the operator's hands are in no danger of being hurt between the mold halves.

Automatic stripping, being a very fast operation, requires a pneumatic (air) pressure system, just as do the blowing proper, the cut-off at the die, the operation of automatic valves in the die head, etc. Because so many moving parts are thus actuated by air, it is sometimes recommended to oil the air in the pneumatic system lightly.

Hydraulic (always oil) systems are generally used for clamping and moving the molds, the platens on which they are mounted, and other heavy parts of the molding equipment (such as rams) which are used in some molding procedures.

Additional Reading

Griep, J., S.P.E. Seminar Presentations. SPE Brookfield CT (1993-1999)

Irwin, C., Johnson Controls, Manchester, MI. Company literature

Irwin, C., *Extrusion Blow Molding Tools* (1985) Center for Professional Advancement, Edison, NJ

Raddatz, E. and Gondo, C., *Hoechst Blow Molding Handbook*. Hoechst AG, Frankfurt, Germany

Rosato, D. (Ed), *Blow Molding Handbook*, 2nd ed. (2003) Hanser Publishers, Munich

Suit, M., *Blow Molds*, In *Modern Plastic Encyclopedia* (1986) McGraw-Hill, New York

A Short Course in Blow Molding High Density Polyethylene (1970) Chevron Chemical Co., Houston, TX

10 Understanding Injection Blow Molds

This section will discuss the following:

- Tooling
- Parison neck ring
- Parison (pre-form) cavity
- Bottom mold neck ring
- Blow mold cavity
- Bottom plug
- Core rod assembly
- Injection manifold systems
- Parison (pre-form) design

10.1 Tooling

Critical tolerances

Tool components for injection blow molding must be machined to precise tolerances because of their critical combined function in the overall molding cycle. The parison molds are more critical than the blow molds since they are subjected to direct hydraulic clamp pressures and material injection pressures. The core rods are subjected to the same conditions as the parison molds and must be concentric to initiate good blowability. The blow molds are subjected to only a third of the clamping force of the parison cavity. Due to the precise machining and interrelations of the component parts necessary to make up a complete injection blow molding tool, excessive dimensional variations between tool components are not tolerated.

10.2 Parison Neck Ring

Forming the shape of the neck

The parison neck ring forms the finished shape of the threaded or neck section of the container (see Fig. 10.1). It also centers and securely retains the core rod inside the parison to alleviate core rod deflection during the injection part of the cycle. The neck ring is a separate component, which must be precisely fitted into the parison cavity to obtain a flash-free part. The parison neck ring has its own independent thermal control channels for conditioning the molded neck finish.

The parison neck ring is made from nondeforming air-hardened tool steel.

Parts of parison mold

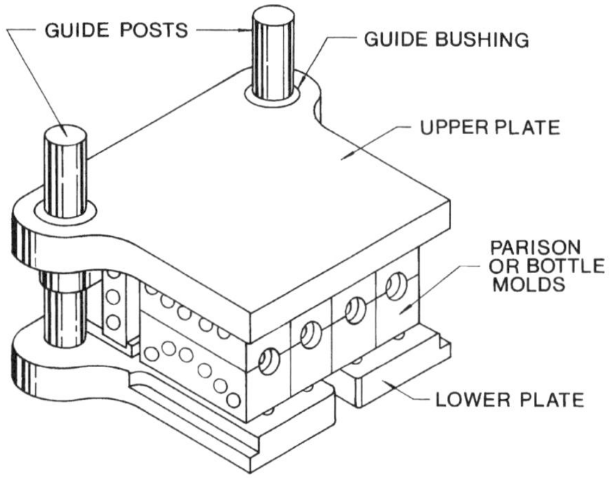

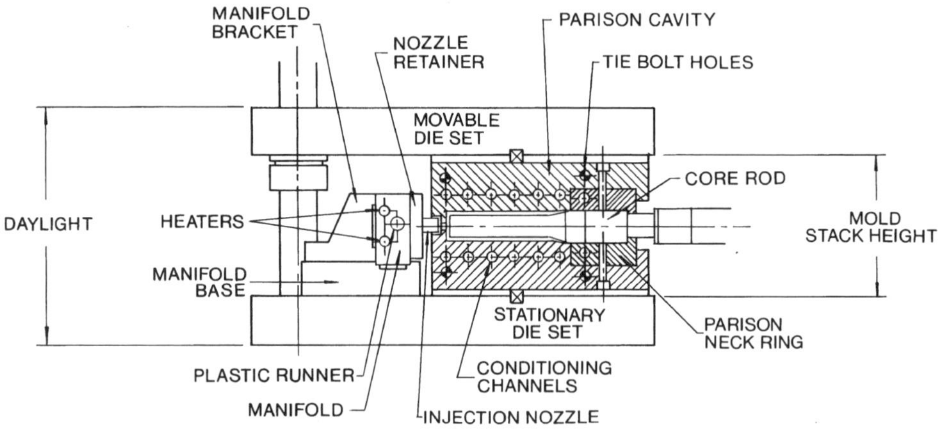

Fig. 10.1: Parison mold assembly

10.2.1 Parison (Pre-Form) Cavity

The parison mold consists of stationary and movable halves, with cavity dimensions determined by the weight, wall thickness, and type of resin of the final part to be injection blow molded. The width of the parison mold is governed by the number of mold cavities that will fit into the specific machine, which is limited by length of the machine trigger bar, the machine clamp tonnage, and the plastifier melt capacity.

Temperature zones critical for uniform blowing of parison wall

Temperature (material conditioning) is an essential part of the cavity design. Temperature zones are created by channels drilled into the mold body for the circulation of a liquid conditioning medium. These channels are connected to form individually controlled temperature zones. The number and location of the zones within the parison cavity are a critical consideration in the design as they directly affect the blowability efficiency of the parison shape. Each temperature control channel is positioned along the parison cavity profile to serve a specific predetermined purpose so that the temperature can be varied along the parison profile to ensure uniform blowing of the parison wall.

Parts of blowing mold

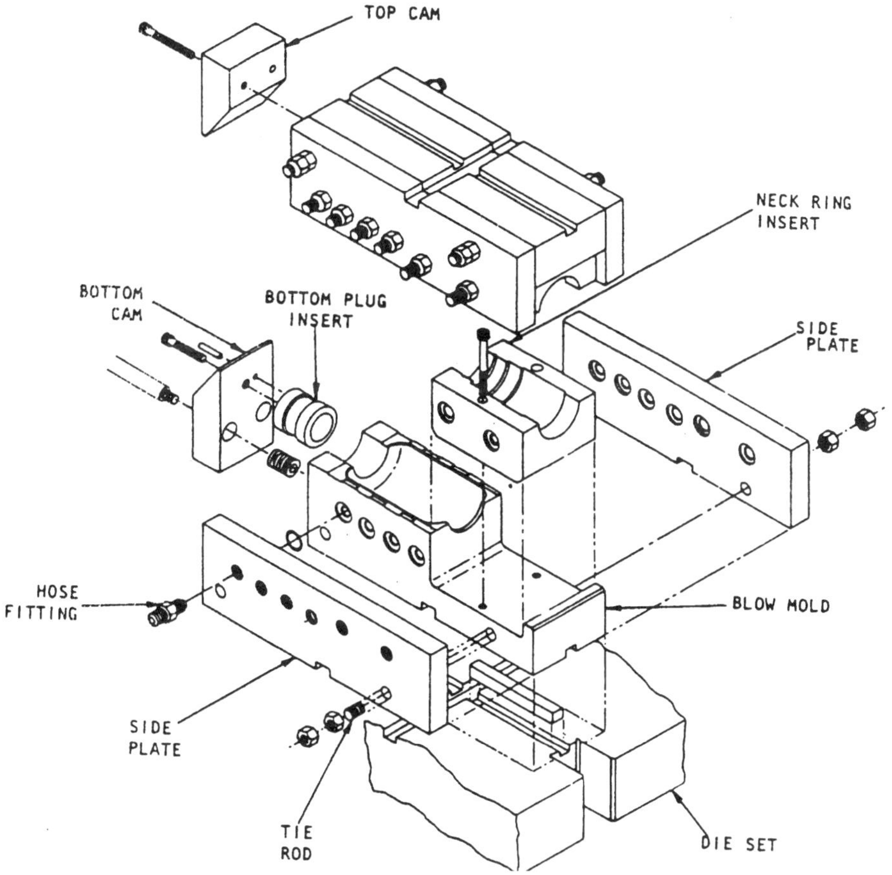

Fig. 10.2: Blow mold assembly

Depending on the type of plastic material being molded, parison cavities are made from air-hardened tool steel or prehardened steel.

10.2.2 Blow Mold Neck Ring

Securing the neck finish

Unlike the parison neck rings that actually form the desired neck finish shape, the blow mold neck contains and secures the already formed neck finish during the blow sequence of the process (see Fig. 10.2).

The neck finish diameter of the blow mold neck ring is slightly larger than the parison neck ring so that it can accept the already formed parison neck finish without distorting it during transfer and seating into the blow mold neck ring.

Independent cooling is incorporated in the blow mold neck ring for thermal control.

Depending on the plastic material being molded, the blow molded neck ring is generally made from air-hardened tool steel or prehardened steel.

10.2.3 Blow Mold Cavity

The blow mold cavity consists of stationary and movable halves similar to the parison mold cavity. Depending on the type of plastic material to be molded, these cavities are made from air-hardened tool steel, prehardened steel, aluminum, or beryllium copper.

Lower blow clamp pressure

Blow molds are used to shape the final form of the container and are not exposed to the severe clamp and injection pressures of the parison mold. The blow mold is subjected to the minimal blow clamp pressure needed to withstand the 1034 kPa (150 psi) air pressure necessary to blow the parison shape.

The most important consideration in the blow mold cavity design is to ensure adequate cooling. Here again, as in the parison mold cavity, peripheral channels should be provided for maximum cooling to help reduce overall cycle time.

10.2.4 Bottom Plug

Forming the bottom configuration of the container

The bottom plug is an independent part of the blow mold cavity and forms the bottom configuration of the container. It is usually made in halves, may be fixed in position or retractable, and may be water cooled. For most polyolefin containers push-up heights can be stripped over the fixed bottom plug when the mold is open. With polystyrene, polycarbonates, and other rigid materials, a retractable bottom mechanism is necessary.

When a retractable bottom mechanism is used, the upward movement of the top half of the mold allows the bottle to evacuate the blow mold cavity without stripping or distortion. The retractable mechanism can be operated by a cam or, in difficult cases, by air or hydraulic cylinders (see Fig. 10.3).

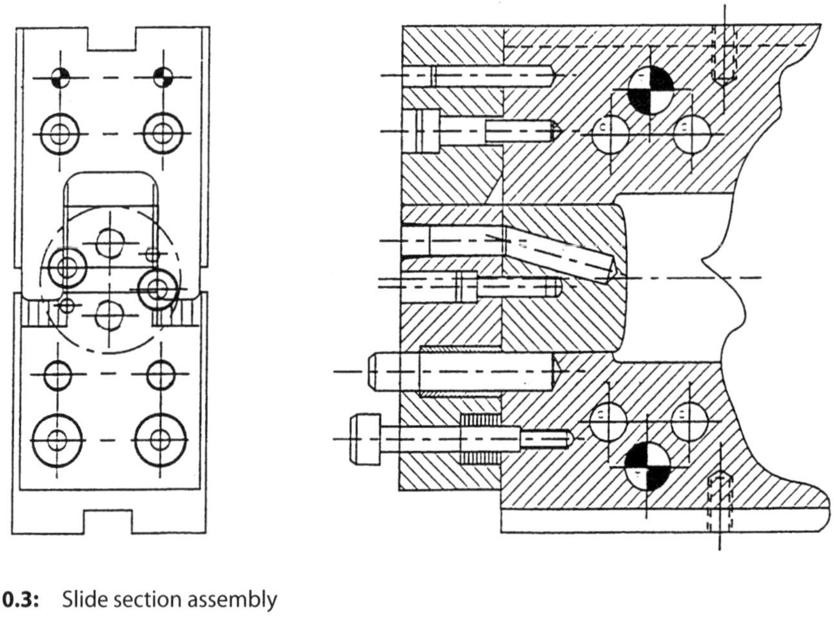

Fig. 10.3: Slide section assembly

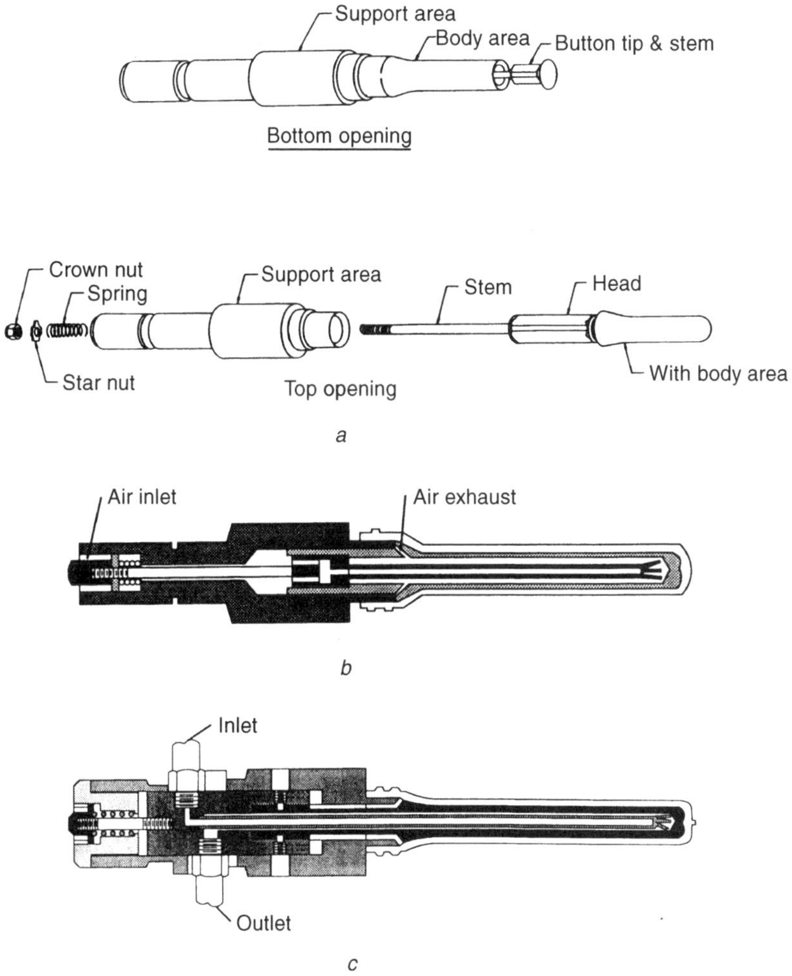

Fig. 10.4: Core rods

10.2.5 Core Rod Assembly

The core rod forms the inside diameter of the neck finish and the inside shape of the parison (see Fig. 10.4). The core rod incorporates the air channel and valves used to blow the plastic melt into the final shape of the bottle. Core rods are usually made of oil-hardened tool steel. A hex nut, spring, and star nut assembly is positioned at the back end section of the core rod valve for blowing and closes the valve prior to injection.

Non-temperature controlled core rods are heated by the melt to provide blowability; however, they do not maintain a consistent temperature profile along the molding surface as do core rods with temperature control. With the advent of heat-sensitive and engineering thermoplastic resins, temperature controlled core rods provide a more accurate and consistent temperature profile along the core rod molding surface.

Providing accurate and consistent temperature profile

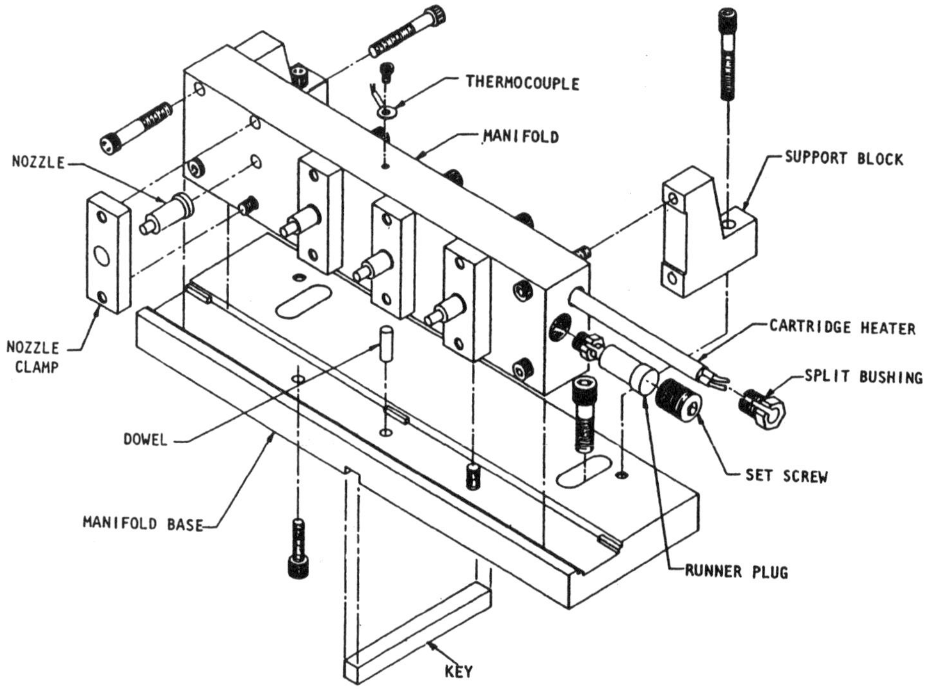

Fig. 10.5: Manifold assembly

10.2.6 Injection Manifold Systems

The manifold assembly is fastened to the stationary parison die set. The manifold assembly itself is made up of the manifold base, nozzle clamps, and nozzles. The injection nozzles are held by the nozzle clamps with retaining screws accessible from the rear of the manifold (see Fig. 10.5).

Simple but adequate design

The hot melt is injected from the machine nozzle into a manifold, which performs the same function as a hot-runner manifold in regular injection molding. The manifold design used for injection blow molding, however, is far less sophisticated than the type used in hot-runner or insulated-runner injection molds. The simple manifold design has been adequate because most resins used for injection blow molding are easily processed. But with the advent of higher-cavitation, heat sensitive, and engineering thermoplastic materials, more sophisticated special manifold designs are required (see Fig. 10.6).

The injection nozzles allow passage of the melt from manifold to parison cavities. Nozzle orifices generally vary in diameter from cavity to cavity in a given mold set to provide a balanced fill of all cavities.

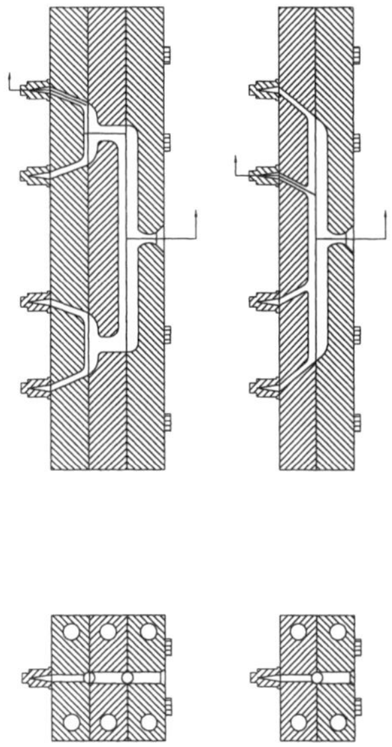

Fig. 10.6: Cross section of injection manifold

10.2.7 Parison (Pre-Form) Design

A crucial element of injection blow molding is the configuration of the core rod and parison. Each container shape has its own unique core rod and parison design. If the initial design of the parison and core rod is not suited for the particular bottle to be molded, a production processing problem will continue until the design is corrected.

Unique design for each container

The designer should always keep in mind maximum processibility or blowability when selecting the parison and core rod shapes. Do not, under any circumstances, sacrifice processibility for maximum cavitation. Consider injection clamp tonnage relative to the projected area of the parison shape, the physical limitation of the machine platen area, the machine trigger bar length, and the total shot capacity of the machine plastifier. Always insist on building a single or unit cavity tool first before going into production tools. Much can be learned about molding characteristics of a particular core rod and parison design through single cavity sampling. If the design is difficult to process during the single-cavity stage you can be sure the process window will be narrow in production, and this is not a desirable goal. Design changes are normally made during the single cavity stage to produce parts in specification with as broad a process window as possible. Normally the unit cavity tool can become part of production tooling.

Single cavity sampling

Additional Reading

Belcher, S. (Chapter 3) and Wunderlich, E. D. (Chapter 24). In *Plastic Blow Molding Handbook*, Lee, N.C. (Ed.). (1990) Van Norstand Reinhold, New York

Lee, N. C., *Blow Molding Design Guide* Chapter 10. (1998) Hanser Publishers, Munich

Albramo, Ralph, S.P.E. High Production Blow Mold Seminars Manual, Brookfield, Connecticut (1993-1999)

11 Finishing

11.1 Part Molding Equipment and Processing

The finishing of a blow-molded part should be considered in the product design, mold engineering, and process planning phases. When actually making a part, imagine the mold opening, the part hanging still in the parison stage, and the flash still attached.

The following potential operations need to be considered:

1. Removing a dome or other sections from the part body
2. Removing the flash and drilling operations
3. Decorations – hot stamping, heat transfer, and serial numbering by the hot stamp method
4. Automatic weigh and record equipment
5. Safety – ergonomics and noise control

11.2 Planning for Finishing a Blow Molded Part

11.2.1 Product Design

Optimize part design

1. Radii, no square corners either inside or outside
2. Wherever possible, includes orientation and register features, for positioning and holding in downstream tasks
3. Lay out flash pockets in the product drawing
   a. Wide enough to accept normal flash variations
   b. Circumferential flash is to be avoided, if at all possible
   c. When circumferential flash is unavoidable, compensate for it in the mold design

11.2.2 Mold Engineering

Optimize mold design

1. Can the part be molded "Siamese?"
2. Is the application suited for "family" molds?

3. Flash pockets
   a. Depth should be calculated; plan to cut metal safe; don't count on the old rule of thumb, 1.8 mm (0.070 in) deep per side
   b. Do keep flash pocket depth tight enough to get cooling contact, so flash is rigid. (Molten wavy flashes, in addition to increasing the risk of burn accidents enlarge the potential for rejects due to parisons and/or parts welding together, and minimize the potential for using deflashing dies.)
   c. Consider corrugating the flash, especially head and tail tabs. (This is more expensive, but provides superior results, and can decrease cycle time.)
   d. Evaluate "blowing the flash," which provides a dimensionally stable, cool flash. (Again, a fringe benefit can be decreased cycles time.)

Optimize mold design

4. Accommodate circumferential flash. Consider employing "flash breaker" pinch-offs, so that flash can be removed in sections, especially important when deflashing is performed by machine. A full ring flash must usually be removed by hand because this is very difficult to automate.
5. Pinch-offs – extremely important
   a. Width – keep to a minimum – 0.381 mm (0.015 in) recommended
   b. Hardness – beryllium copper, preferably, or steel, preferred over aluminum, for long life and maintaining sharpness
   c. Incorporate in-machine replaceable inserts, whenever and wherever possible
6. Part ejection – configure the mold to eject the part so it clears the mold without human intervention
   a. Utilize re-pinch mechanisms as strippers
   b. Plan to include mechanisms to retain the mold to a preferred mold half, then include ejection features (see chapter 9).

11.2.3 Process Planning

1. Lay out the entire process, step by step.
   (What has to be done when, and where to do it?)

Location

Whenever possible, finishing the part to the maximum degree possible, possibly even the packaging of it, in the molding room, adjacent to the blow molding machine, promises the best economics, as it eliminates the labor involved in warehousing, transport, and storage. An additional benefit is that the chain of communications for the purpose of quality assurance, back to the molding machine, is short and timely.

The following solutions should be considered:

a. Labeling
b. Decorating, i.e., laminating a finish membrane to the part
c. Drilled holes
   - In compression molded areas, consider punching in the mold.
   - In blown areas, consider a bubble which can be snare cut, rather than drilled – chipless!

d. Threads — Threads
   - Internal threads can be accomplished by use of unscrewing cores.
   - Threaded cavity sections along parting lines can form external threads.

e. Deflashing — Flash
   - In entirety
   - Partial, in stages

2. Investigate making the flash productive during its limited life. Can it be configured to use as a temporary handling, positioning, or registration device?
3. Part handling – removal; if this can be automatic, which is preferred, it enables gate closed constant operation. — Removal
   a. Stripper "pop and drop"
   b. Via side takeout
   c. Via bottom takeout
4. Manual or operator controlled trimming and deflashing, unless the operator is exceptional, the cycle varies shot to shot, creating inconsistencies in temperature history, which creates shot to shot variation in product weight, location of programming points, and dimensional variations. This is demonstrated by the readily visible symptom of tail length, which is actually due to overall parison length variation. The operator will tell you "those tails keep changing on me." Actually the operator is changing the tail on you. — Trimming
5. Parts handling – secondary cooling — Handling and cooling

   Parts up to 2.286 mm (0.090 in), if molded in a wall-cooled mold with reasonably designed flash pockets should not require secondary cooling. Medium wall parts sometimes are conveyed through fan cooling. Heavy wall parts – 9.525 mm (0.375 in) and heavier – may be passed through a chilled water spray, or conveyed through a chilled water bath.

   Holding the parts until they cool, prior to deflashing, is more often than not due to the retained heat in the flash. Deflashing early, rather than later, minimizes the hazard of the retained heat migrating back into the part. Thus early flash removal decreases the possibility of warpage.
6. Parts handling – warp control or shrink fixturing — Shrinkage

   Not all parts will require shrink control. Depending on part shape the part must be placed in shrink fixtures for whatever amount of time is necessary. Thicker parts tend to require more time in the fixture than thinner parts. There are no known rules of thumb; experimentation with the actual part is the norm. Shrink fixture dwell time is usually stated on the shop floor by number of shots.

   The control of shrink fixture dwell time can be handled on a shrink table, with manually operated clamps. The operator keeps track of where the coolest part is, replacing it with a part fresh from the mold. This method is the least desirable.
   a. Carousel or turntable shrink fixture. Operator indexes the table each shot, removing and replacing parts in rotation
   b. Indexing conveyor, as part of an automatic finishing line
7. Parts handling – conveying — Conveying

   The first consideration is whether the part to convey has been deflashed or not. Deflashed parts generally can be moved with simple, single speed, constant running belt conveyors without problems.

Conveying parisons fresh from the mold at the same time as parts is a process that tends to be reject-prone, due to the hot flash sticking to other parisons, or worse yet, parts. When conveying fresh from the mold parisons with parts is required, depending on flash configuration, conveyors with support rails for the flash are recommended. When parts are circumferentially flashed, conveyors with separators, guaranteeing that flash from one part cannot touch an adjacent part or its flash, are required.

Deflashing

8. Trimming and deflashing

Trimming and deflashing are reasonably synonymous in industrial settings, and in the large part blow molding industry, meaning the removal of flash. (Trimming is more descriptive in container blow molding, where blow domes are cut away.) The most common method of flash removal is manual, performed with the state of the art finishing tools we looked at previously.

Increasingly, machine deflashing is being employed with an assortment of pneumatic and hydraulic deflashing presses. For the most part, they are semiautomatic, and operator loaded and unloaded. The majority are simple designs, with nests mounted on tables. A few have channels for the scrap to drop to a conveyor below. This facility is negated when deflashing parts with circumferential flash.

a. Common deflash press tooling of female bottom nests, in which the part is inserted
b. Where interior flash webs must be removed, the bottom nest includes a die or cutout; the top tooling includes a punch

Deflashing presses may be either bottom or top acting. The part cradled in the nest contacts the deflashing die, and over strokes slightly. A top acting stabilizer/ejector member is generally required.

Tooling for head and tail flash, as well as for circumferential flash, will vary depending on whether the parting line is flat or irregular. Flat parting lines are well handled by a die plate. Irregular parting lines are better handled by opposed chisel tooling.

Fixtures

There is a category of secondary equipment referred to as fixtures. Generally fixtures are job specific for a specific part, designed for secondary machining. Some fixtures include deflashing, both partial, i.e., an interior section which needs to be punched out, and complete exterior deflashing.

Guillotines

A feature common in deflashing machinery is guillotines, which chop off blow channels and neck-like areas, see Fig. 11.1.

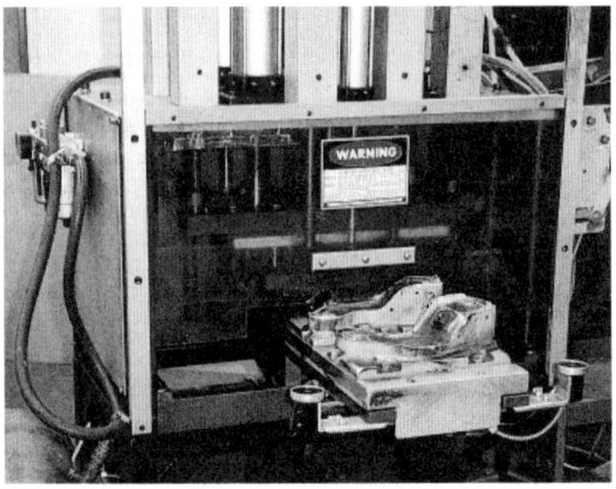

Fig. 11.1: Guillotine deflashing fixture with sliding nests. Courtesy WEK South Corp., Reidsville, N.C.

11.3 Removing Domes and Other Sections

11.3.1 Rotating Saw Blades

Common trimming tools

Often rotating saw blades are used to remove domes or other sections from the part body. Disadvantages are:

1. Very noisy
2. Clips and fluff made by the saw blades when cutting the part create a mess in the work area and can affect air quality for the workers
3. Danger from exposed saw blades and chips being flung that may get into the eyes
4. When hydraulic and electric motors are used, chips can get into these motors and cause them to burn out quickly
5. Finished cut not very smooth and requires extra trimming

11.4 Knife Cut

The better way is to use a knife cut when possible. Figure 11.2 shows a groove for a knife cut. The groove size for a knife cut will depend on the size of the part and the thickness of the wall. Generally the height and depth of groove should be at least twice the wall thickness with shape corners to prevent the knife from wandering.

Groove design for knife cut

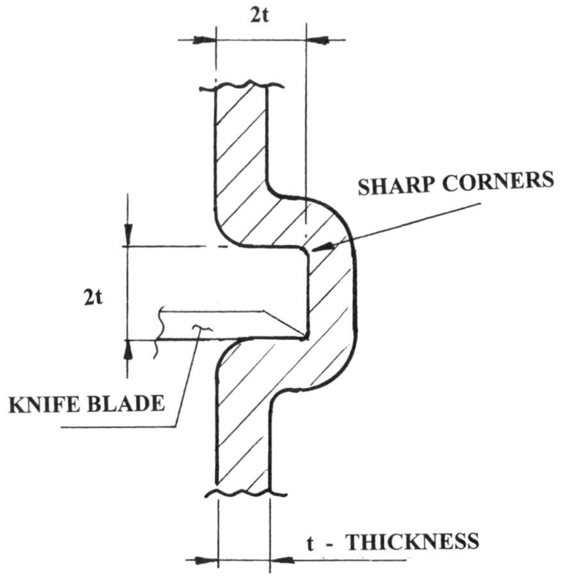

Fig. 11.2: Groove for knife cut

11.4.1 The Cutting Machine – Round Parts vs. Parts with Corners

Round parts can typically be cut in a machine which rotates the part; the blade is brought in manually, see Figs. 11.3 (A), (B) and 11.4.

When squared parts are rotated, some type of control is needed to maintain the depth of the blade into the part, to automatically bring the knife in and out, and to start and stop the machine in the same orientation each time. Other operations can also be added to the auto cut machine if needed.

(A)

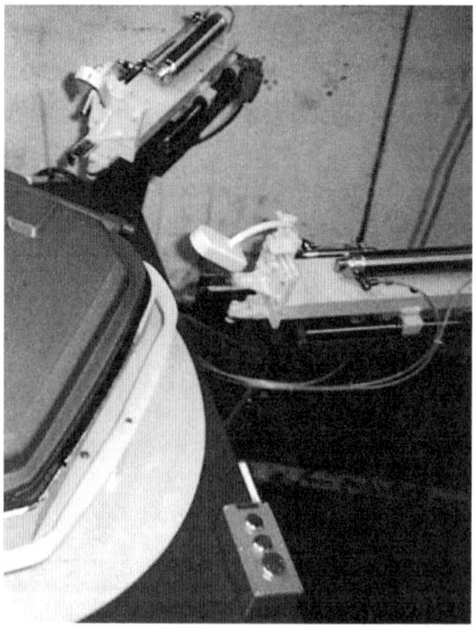

(B)

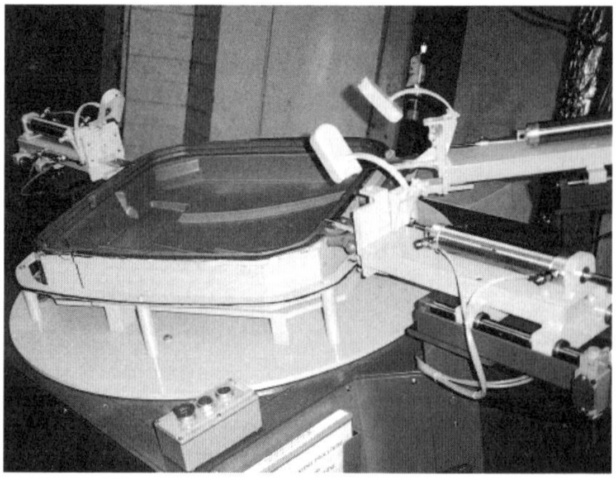

Fig. 11.3: (A) Knife cutting machine – before cut; (B) Knife cutting machine – after cut
Courtesy WEK South Corp., Reidville, N.C.

Fig. 11.4: Knife cutting machine – double cut

11.4.2 Benefits of Using a Knife Cut

Advantages of knife cuts

1. No noise
2. No chips
3. No danger from chips in eyes or being breathed into lungs
4. Eliminates chips from going into motors and burning them out
5. Knife cut leaves a clean, finished cut
6. Cosmetics
7. No saw blades to replace, order, or keep in stock
8. Greatly reduces maintenance and downtime
9. Knife blades can be replaced without downtime and are often so cheap they can be thrown away
10. Cleanup in work area is greatly reduced
11. Workers are pleased with cleaner work environment
12. Reduces rejects by providing more accurate cut
    a. Savings of $200,000 per year are possible
    b. OSHA compliance – noise, air quality, and guarding against accidents
    c. No saw blades to buy, resharpen, or stock
    d. Reduced maintenance labor and cleanup labor
    e. Less employee turnover
    f. Greatly reduced hydraulic and electric motor replacement

Safety

Efficiency

Cost effective

11.5 Other Methods

Making an appearance at a growing rate are integrated finishing lines, which include part handling and deflashing as well as other secondary operations. Although few in number to date, articulating and multiaxis robots tooled with water jet cutters or router bits are slowly being brought into the blow molding industry. On the pro side they deflash and deburr in one operation. Unfortunately, speeds appear to be low at only 1.016 to 1.524 cm/s (4 to 6 in/s) in the units I have viewed in operation. Multiple robots employed on a single part to increase production rates do not seem practical, except on very large parts.

11.6 Deburring

This term refers to the removal of the flash extension at the parting line. The parting line remains when the flash is removed. In most parts, sharp extensions are not permitted and must be removed.

11.6.1 Common Deburring Methods

a. Hand scraping, using a paint scraper or some simple deburring tool borrowed from the metalworking industry.
b. Flame polishing in which simple parts are conveyed past a gas flame. Handheld torches are also used to blunt sharp edges when the flash extension is short and narrow.
c. Belt and orbital sanders can be employed for parting lines in convex curved surfaces.
d. Routers handle parting lines in a variety of planes. The drawbacks to routers are noise and fine debris. For routing operations to be successful, adequate jigs and fixtures are required to achieve quality results. A large diameter cutter is recommended over a typical router bit. Usually for a large diameter cutter, the debris generated is more of a chip form than a fine particle, thus the debris is easier to control, collect, and reclaim.

Avoid static cling of debris

On the subject of chips and debris, consider using a flow of low velocity deionized air in the cutting tool area to minimize the static cling of the debris.

11.6.2 Manual Flash Removal

Manual flash removal, employing a handheld knife, varies in quality of trim. Depending on the personal mechanical skills and the repeatability of the person handling the knife, quite often varying length flash extensions are left, which then have to be removed.

11.6.3 Machine Flash Removal

Machine flash removal does not cut the flash away. Instead, the flash is fractured or sheared. Neither the deflashing machine nor the deflashing die control the amount of flash extension. Instead the amount of remaining flash extension to be removed is dependent upon:

a. Mold pinch-offs, width, condition
b. Mold closure, affected by:
   - Whether flash protrudes beyond flash pockets
   - Flash pocket depth adequate for volume of flash
   - Adequate pinch clamp force
   - Platen/clamp distortion

Try to minimize/ eliminate deburring

Deburring is often expensive in terms of time, money, and high injury exposure. Any effort to minimize and/or eliminate deburring should pay dividends.

11.7 Clamping Needs in Extrusion Blow Molding

The major factors affecting the two most common post mold operations in extrusion blow molding, for both container and industrial and large part blow molding, occur when the mold closes and clamps up. Blow molding is a low force process compared to other plastic processes, such as injection molding, or other manufacturing operations, such as die casting and metal stamping. Blow molding professionals generally do not pay attention to this particular aspect of our process to the degree an injection molding professional does, for example. In an injection mold, if feather flash is being generated at the parting line of a part, deburring is not considered as a secondary operation. Instead, three items are examined to eliminate the cause of the flash: the process, the mold, and then the machine.

Two clamping requirements

In summary, there are two clamping needs in extrusion blow molding: pinch clamp force and blow clamp force. For the most part, we concern ourselves with:

a. Will the mold fit in the clamp?
b. Will the part come out when the mold is open?
c. Will the mold stay shut when the blow air comes on? (Enough blow clamp force?)
   See Chapter 7.

11.7.1 Blow Clamp Force

Blow clamp force is a force which keeps the mold closed when the part is inflated and full blow pressure achieved. With most recent blow molding machines manufactured in the last ten years where the mold fits in the clamp, most probably there is enough clamp force to keep the clamp closed. This assumes that the accumulators are working and keeping the clamp from breathing when another actuator needs hydraulic flow and a momentary system pressure spike occurs.

Problems when clamp does not close properly

However, once the blow air comes on, it is too late to change the blow clamp force. Whether or not the clamp is adequately closed, the part is a done deal. Either the clamp is closed or it isn't. If it did not adequately close, then the damage is done, and the operator is faced with paying the price in additional and unnecessary finishing labor. The concern is the web of flash at the pinch-off. The thicker the pinch-off web is, the more force is needed to deflash the part. A thicker web generates a more jagged and sharper flash extension.

11.7.2 Pinch Clamp Force

Pinch clamp force is the force required to close on the parison, compressing it between the pinch-offs, and deforming it into the flash pockets.

The need for pinch clamp force decreases once the mold has been closed as tight as it is going to close. The blow air comes in to inflate the mold, and the requirement changes to blow clamp force. The need for these respective forces occurs at different times. They are sequential. They are not cumulative.

The sequence is:

a. Mold closes requiring
   - Pinch clamp force
b. Mold closed signals blow air on, requiring
   - Blow clamp force

Determine clamp force

How can the requirements for pinch clamp force be determined?

1. Measure the total length of pinch-offs in the mold.
2. Multiply the linear measurement of pinch by a rule of thumb for the resin that is being molded.

Through experience and experimentation, the following rules of thumb for pinch clamp force have been determined. Some of this material appeared in a paper presented at the S.P.E. 1975 ANTEC. Further mention was made in the S.P.E. 1991 Blow Molding RETEC paper. These rules of thumb have stood the test of time.

Resin	Force Required per Linear Centimeter (Inch)
HDPE	700.507 N/cm (400 lb/in)
PP Homopolymer	866.878 N/cm (495 lb/in)
PP Copolymer	954.441 N/cm (545 lb/in)

Rules of thumb

These rules of thumb are based on the use of typical extrusion blow molding grades of resin in molds having 0.004 cm (0.015 in) pinch-off lands.

If wider pinch-off lands are used, increase force as shown:

- 0.762 mm (0.030 in) land width – add 10 percent to required force
- 1.143 mm (0.045 in) land width – add 20 percent to required force

11.8 Effects on the Force Required to Deflash a Part

Typically a deflashing machine will exert a uniform force at the same velocity, cycle after cycle. If the parts are inconsistent and vary in web thickness, machine deflashing will reflect the variations.

The flash extensions will vary depending on how well the mold has closed. Thus, the deburring effort, time, and ultimately cost will all be affected.

Determine deflashing force

Through some basic and simple research, combined with practical experience, some rules of thumb for deflashing force have been determined. The initial objective was to be able to size deflashing press force based on fact rather than guesswork. In querying contacts in the industry, it was found there was no rhyme or reason to the sizing methods. If x amount of force was not enough, try 2x!

There are standards or ASTM tests normally used on plastic resins that would enable us to hypothesize a method. Reaching into the metalworking field, and looking at how shear rate is calculated, some correlation with the ASTM test method D638, tensile yield, psi, and tensile strength at break was found. The metalworkers say "do the calculations and then add a 50 percent safety factor." The difference between theory and experiment led us to choose a multiplier of 2.5 when dealing with practical deflashing applications.

For example:

Example

- Conditions
- Mold-new, perfect as possible
- Pinch land width 0.381 mm (0.015 in) (BeCu)
- Machine-new, clamp square within 0.0508 mm (0.002 in)
- Hydraulic clamp
- Material: HDPE

To achieve 0.0762 mm (0.003 in) web at the pinch-off between the part and the flash, check the deflashing force required, coming up with a figure of 155.688 kg/cm 35 lb per lineal in, using a hydraulically actuated deflashing press.

By shimming the mold faces, we deliberately produced sample parts with increased web thickness at 0.0762 mm (0.003 in) increments. This finding indicates that for every 0.0762 mm (0.003 in) increase, an additional 155.688 kg/cm (35 lb) of force needs to be added. In the field, testing showed that machine and process variation in the real world was of an extent such that for a deflasher to work well, in a reasonably controlled plant, forces were required to be at about 17.858 kg/cm (100 lb per lineal in). If that does not do it, then the part needs to be examined, and the process, mold, or machine corrected.

11.9 Finishing by Machining

Machining operations include most of the typical processes and methods used for nonplastic materials:

Machining methods

- Drilling
- Sawing
- Band sawing
- Radial processing
- Routing
- Milling
- Reaming
- Facing

For the most part, perishable tooling used in the metalworking and woodworking industries are adapted. While our materials are softer, they are gummy, increasingly so if we are loading them enough to create frictional heat. At a point where the frictional heat builds sufficiently, cutting tools begin to slip. Withdrawn from the workpiece, it will be observed that the resin has formed into a solid mass, blocking the cutting face from the surface to be cut.

11.9.1 Drilling

Best practice

Drilling is an easy to describe example. The deeper the hole, in relation to the diameter, the greater the tendency for chips to begin to melt. 1:1 ratios of diameter to depth are no problem. However, when a ratio greater than 1:3 is needed, withdrawing the drill and clearing away the chips every other diameter will usually provide a cleaner hole quicker than drilling without withdrawing. Unnecessarily high tool speeds are counterproductive.

Work-holding remains more of an art than a science to a large degree, but it is a challenging factor in economical, reject free machining of blow molded products. Our industry still tends to work with simple jigs and fixtures and multiple setups. It is not uncommon to see several drill presses or to have a part taken from station to station, possibly moving the part by hand to use the same tool bit multiple times. High scrap rates, such as 15 percent and more, are not unusual.

11.10 Fixtures

Minimize scrap rates

The next step in work-holding is a fixture, a job-specific device which in most cases fulfills the definition of a machine. An operator places the part once, then activates multiple tools, quite often air powered, with controlled feed rates. When the cycle is complete, the operator removes the workpiece. The scrap rates with fixtures generally are lower.

In some market areas, for example, automotive, the traditional expectation was that the customer paid for and owned the fixture, as well as the mold. That picture is changing. The buyer is now saying, "Those fixtures are machines. We will pay for the special tooling that touches the part, but the molder will pay for and own the fixture." The customer will pay for the drill bit, but not the actuator and chuck. The buyer will pay for the special clamp jaws, but not the clamp itself.

11.11 Computerized Numerical Control

Need for automated equipment

The changing scenario, combined with other pressures upon the industry, is spawning a need for more versatile semiautomatic and automatic equipment developed for the plastic blow molding industry segment. Though computerized numerical control (CNC) machining centers with tool changers commonly used in machining of metal products could be employed, they are built for much heavier loads and duty cycles than our industry needs. In most cases, using CNC, especially considering its high price tag, compares to using a semi-trailer rig to haul a load that barely taxes a pickup truck. CNC machining centers are prime machines in the metalworking industry, while the blow molding professional's need is for secondary machines, lighter construction, and for the equipment to be less expensive.

CNC is increasingly appearing in even smaller metalworking job shops to stay competitive.

Blow molders, while recognizing the need, have not had a simple and easy path to incorporate CNC in their arsenal of secondary tooling know-how and capability. Most blow molding plants at this time do not have people trained to handle CNC equipment, especially on the off shifts.

Much as pencil draftings, and people capable of and willing to do it, are becoming rarities, replaced by CAD, the old time machinists, who could take a sketch or drawing, figure out what had to be done, and how to do it, are also going the way of the dinosaurs. It appears that it is only a matter of time until blow molding shops will have CNC-trained people on staff, moving in from the maintenance and machine shop areas.

Plastic processors, and specifically blow molders, face a challenge nonexistent in the metalworking industry in employing CNC. Part drawings for metal parts can be compared to the part as it is being made, and as soon as it is finished. Metals usually do not change size as they are being machined, while the freshly blow molded part shrinks as it cools.

Dimensional inspection after conditioning

The blow molder, ideally, wants to get the part out of the mold, finished, and done as rapidly as possible. However, for exact dimensional inspection, the part will be compared to the product drawings 48 h after molding, and after conditioning 12 h at 25 °C (77 °F). But in the time frame that it is economically convenient and expedient for the blow molder to machine the part, i.e., immediately upon removal from the mold, the rate of shrinkage is at its peak.

Although the metalworking CNC machinist can, via readily available commercial computer programs, convert a CAD-part drawing to the “G” codes needed to form the tool paths, the blow molder cannot directly convert a CAD product drawing to accurate cutting paths. The reason is that the product drawing provides dimensions after shrinkage is complete. If the blow molder were to machine to the product drawing, the holes would be out of place and the cuts would be too deep.

The idea has been proposed that the tool wear correction facilities of CNC programs could be used to compensate for thermal changes. Unfortunately, this cannot be done, at least not expediently, because blow molded parts do not exhibit uniform shrinkage patterns. Generally, there is greater shrinkage in the hoop direction than in the parison flow direction. Although this is not an impossibility for a competent CNC programmer, it is immensely time-consuming. The blow molder works constantly to decrease the changeover time from job to job. At this point in time, it appears that the task of adjusting tool path programs will neither fit in an acceptable time frame nor be available on off shifts. This challenge awaits an answer, and will be an important link to enabling blow molders to vastly improve their capabilities and economics in the machining operations involved in finishing the product.

11.12 Robotics

The use of robotics slowly increases. A larger percentage of new industrial blow molders ship from the factory with simple robotic takeouts, either side or bottom takeouts, each year.

A few articulating robots are being added, primarily in routine applications. We have seen a few used in surface treatment, e.g., vibratory sanding of automotive parts blown from engineering resins. Interestingly enough, these particular robots were bought for another purpose, then recycled for use in blow molding.

The same plant in the previous example uses robots for spray painting as well. The blow molded parts are removed from the molds, deflashed, then racked. The semi-finished parts are moved to storage, then to a finishing department. The programming care and upkeep of the robots is assigned to specially assigned technicians, and is not part of the blow molding processing group. This department, and its personnel, do not identify with blow molding.

Unlike injection molding machines, blow molders do not have a bank of readily commercial sources which have off-the-shelf designs, which require only an application specific "hand" to accept parts from the takeouts. The reason appears to be that the blow molding industry is small by comparison to injection molding; for example, blow molding machines have less similarity between makes and models, plus blow molders generally do not have access from the top of the molds. Blow molding is still at a point where every application is more or less a special.

11.13 Decoration Equipment and Processing

Generally, decorating means applying a coating to all or specific portions of the workpiece. Typical decorations include:

Typical decorations

- Labels
- Decals
- Screen printing
- Hot stamping
- Pad printing
- Painting

These processes are fields of their own, with numerous options and do's and don'ts, too many to cover in a short time with generalizations and rules of thumb. The best source of information is the vendors, but be specific in the description of the application.

11.13.1 Surface Treatment

Preparing the surface

A requisite step in most decoration is surface treatment. The purpose is to enable the coating, whether it is such items as glue, ink, paint, or plating, to stick to the molded part. Surface treatment involves changing the surface tension of areas to which decoration is to be applied, so that it is higher than the surface tension of the material to be applied.

The majority of blow molded products are still olefins, polyethylenes and polypropylenes. Typically flame treatment is used, exposing the surface to a gas flame. An alternate is corona treatment, in which the part is subjected to corona discharge. More common in container blow molding, corona treatment is not widely used in large part molding.

Surface treated parts, especially those that are flame and corona treated, can be tested via a number of methods, the most simple being a water test, which is simply dipping the part in water. A well-treated surface will wet out with a smooth layer of water. On an untreated or

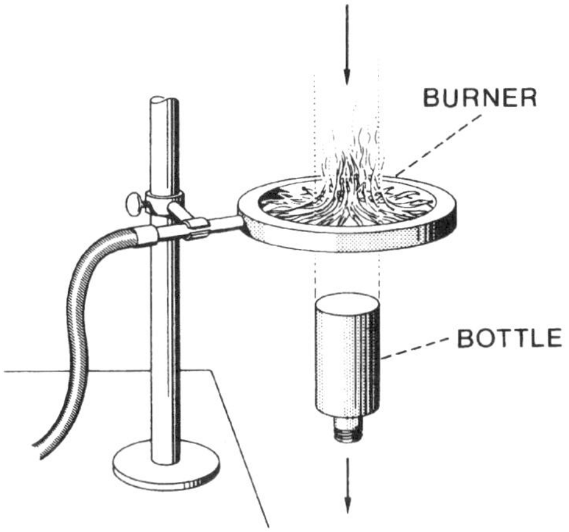

Fig. 11.5: Ring burner for round bottle treatment

poorly treated surface, the water will bead. Water on an untreated surface will remind you of how the water lays on a freshly waxed car. An equally simple, but more discerning test can be performed with commercially available marking pens whose inks are available in differing ranges of surface tension.

Beyond the olefins, other techniques used to surface treat include:

- Washing with water based chemicals
- Solvent cleaning and etching
- Mechanical abrasion-sanding
- Chemical etching
- Additive compounds the resin

Surface treatment is critical

Surface treatment is a very critical step. With applications that you have no experience with, the soundest approach is to query the vendor, the resin supplier, and the supplier of the surface material to be applied. In most cases, they can assist in doing the job adequately the first time, see example of gas flame methods Figs. 11.5, 11.6, 11.7, and 11.8.

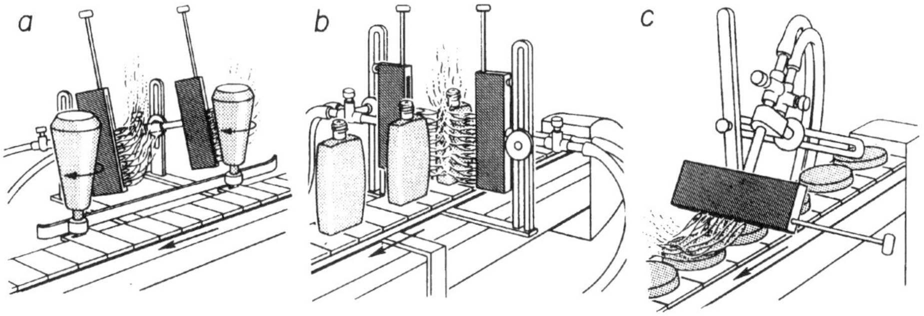

Fig. 11.6: Ribbon burners for a variety of shapes & sizes

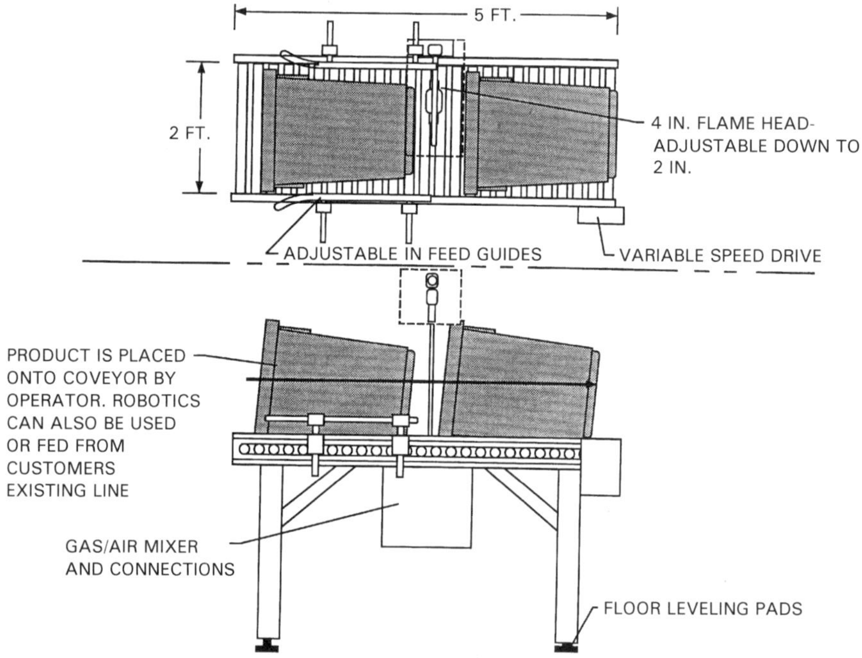

Fig. 11.7: Flame treating trashcans

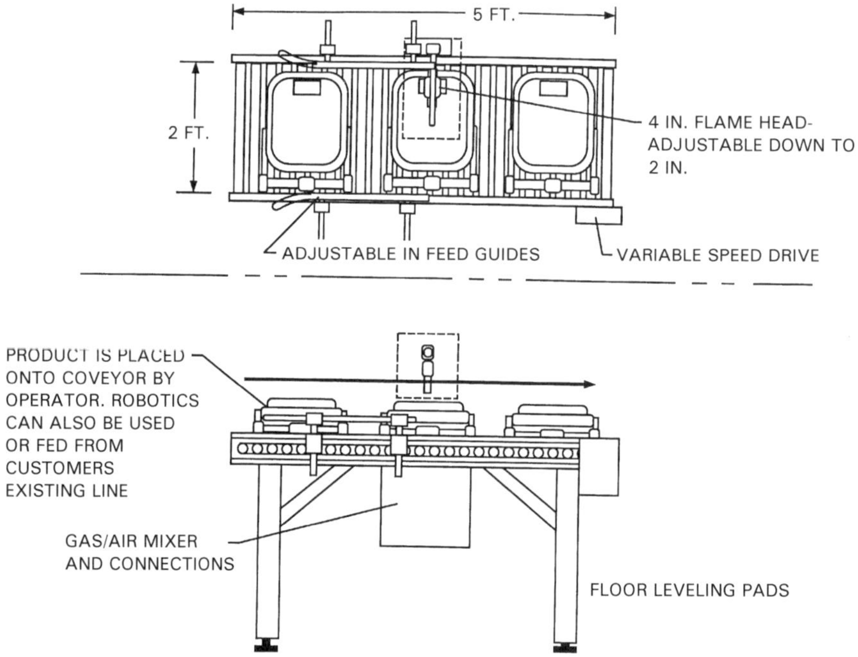

Fig. 11.8: Flame treating trashcan lids

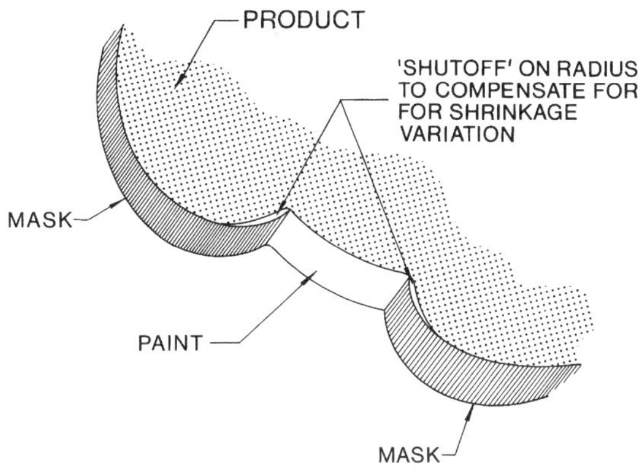

Fig. 11.9: Paint spray mask

11.13.2 Spray Painting

Air atomization and airless sprays are the most common. Methods used by blow molders tend to follow those used in other fields. The process is excellent for applications involving large areas, irregular surfaces, or multiple separated surfaces in a part. Masking is required to separate areas. By far the simplest method of masking involves employing tape and paper to cover areas of the plastic surface. Masking is used for production of automobile bumpers where high quality is required. The tape must be resistant to the solvents used in the spray paint and be able to withstand the drying conditions, and be separated cleanly on removal. The sharpness along the edge of the tape depends on the tape quality.

Excellent for large areas or irregular surfaces

Hard masks are reusable, made from metal or plastic, and placed over the object to be painted to shield paint from selected areas on the surface. Among materials used for plastic masks are polyurethanes and silicone rubber. Metal masks are usually either fabricated by standard mechanical methods or electroformed using copper or nickel. These masks are used, and then cleaned repeatedly in a cleansing solvent tank. The masks are usually duplicated or triplicated to allow cleaning while others are in use. A typical spray mask cross-section is shown in Fig. 11.9.

11.13.3 Hot Stamping

This process involves transferring a coating from a carrier (ribbon or tape), to the receiving surface by pressing a hot die with the image to be transferred against the part. For each image the tape must be advanced. Rotary presses with cylindrical and flat dies are also shown in Fig. 11.10. This process is usually employed in small areas such as printing and logos. It is a time, temperature, and pressure dependent process. A typical application would be the logo on municipal roll out refuse carts, see Fig. 11.11.

Employed for small areas

Fig. 11.10: Host stamp examples on municipal roll out carts. Courtesy WEK South Corp., Reidsville, N.C.

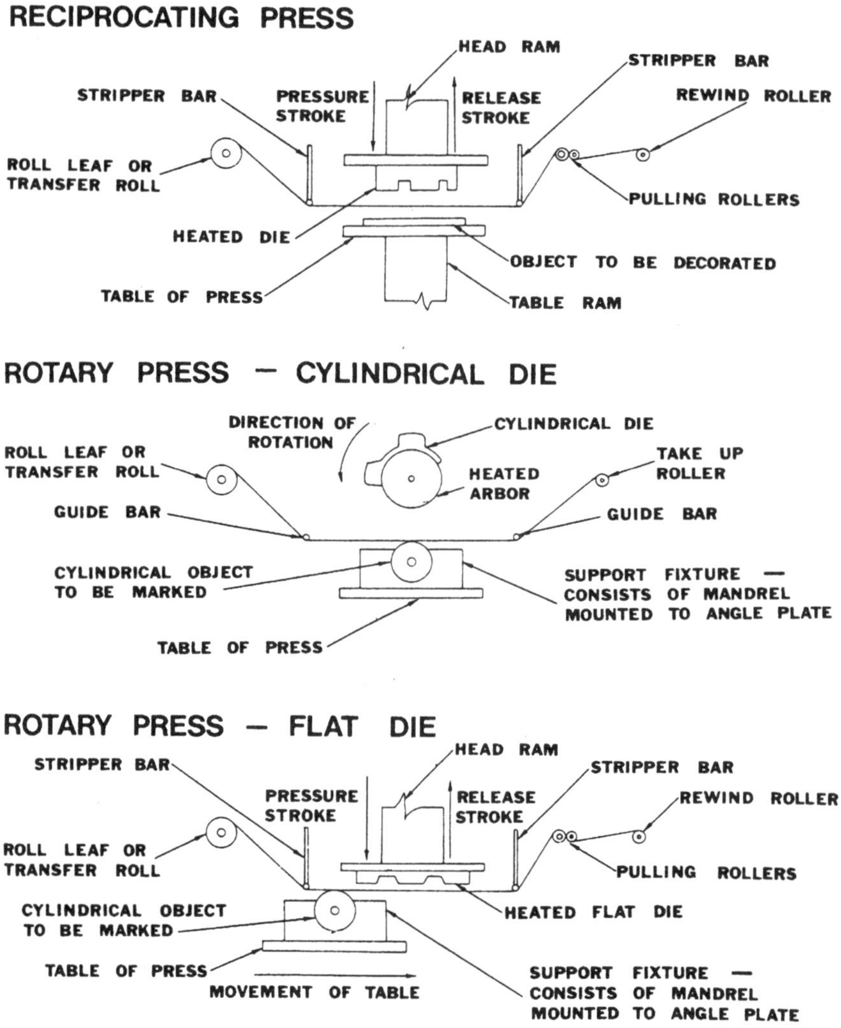

Fig. 11.11: Hot stamping process

11.13.4 Pad Printing

This process permits printing of irregularly shaped surfaces, generally 2.032 × 2.032 cm (0.8 × 0.8 in) maximum. It requires the workpiece to be precisely located and sufficiently rigid to withstand the forces used in the process.

Employed for irregular shaped surfaces

The image is engraved into a hard plate, called a cliché. Cliché materials range from copper, zinc, and magnesium to steel. Photoetched nylon and polyethylene plates are also employed. The volume of impressions usually determines the choice of plate material, determining the hardness of the material as it relates to plate life. The ink is applied to the cliché. A doctor blade is passed across the cliché, spreading the ink evenly, and removing it from the surfaces outside the engraved cavities. A soft silicone pad is compressed on the cliché and picks up the ink on the surface. The soft silicone pad is indexed to the point of printing, then compressed on the workpiece, transferring the ink. As the soft silicone pad's surface is distorted during the compression process, the engraving is also distorted, enabling the transferred image to be accurate. The stages and basic pad shapes are illustrated in Figs. 11.11 and 11.12. Pad printing equipment is more expensive than that for stamping or screen printing (see Fig. 11.14).

Expensive

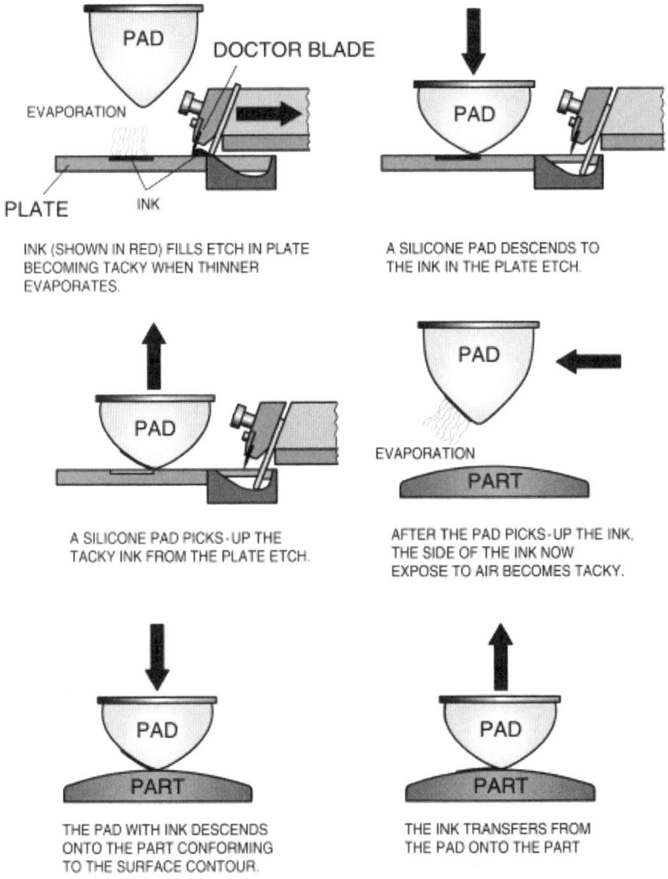

Fig. 11.12: Stages of pad printing

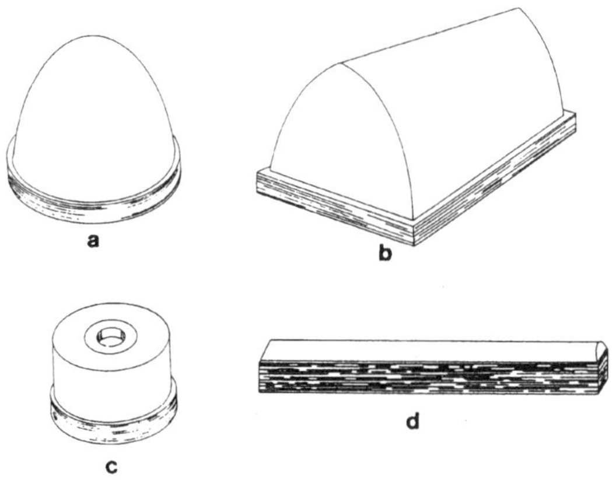

Fig. 11.13: Basic pad shapes: (a) standard; (b) wedge; (c) doughnut; (d) ribbon. Courtesy of Midwest Technical Service.

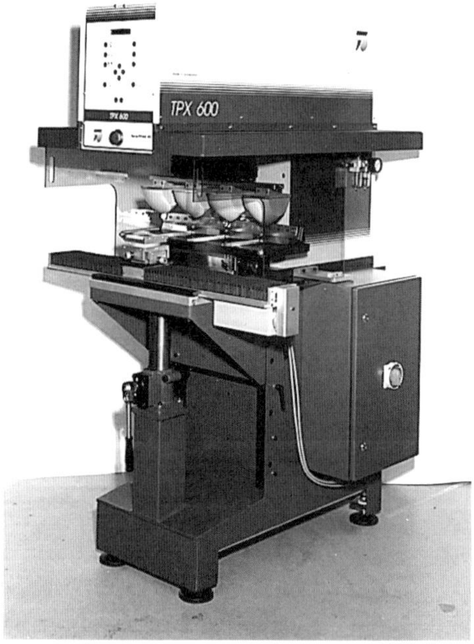

Fig. 11.14: Four station pad printer

Large labels have split stock to enable them to be removed easily by hand, see Fig. 11.15, while automatic equipment is used for small labels. Labels are secured by wet glue or hot melt adhesive application at the time of label application. However, for blow molded products (usually polyethylene) pressure-sensitive adhesive preapplied to label stock is more common. When release paper is run over with a sharp edge, the label disbonds and continues to move on the straight pass. This is illustrated in Fig. 11.16.

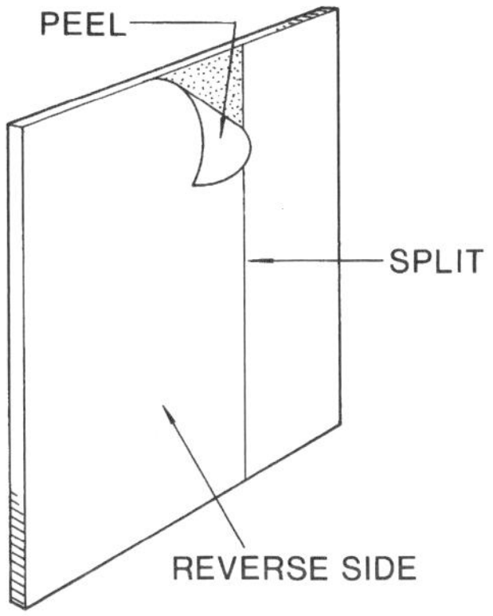

Fig. 11.15: Peel-off label

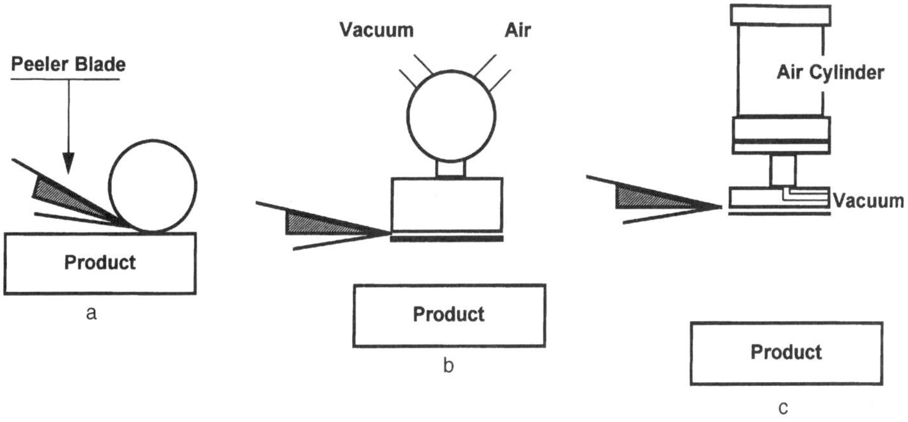

Fig. 11.16: Pressure sensitive label application methods

11.13.5 Screen Printing

Screen printing is a special process in which ink is forced through open areas of the screen. The print pattern is determined by open areas of the screen. This principle is illustrated in Fig. 11.17. A squeegee forces ink through the screen openings which merge to provide a continuous coating over the desired area (Fig. 11.18). The process allows heavy ink coatings and is not possible in other printing methods. This process is simple and versatile; therefore the equipment is expensive. It is slow, however, and it is not generally used in high production operations.

Used in high production operations

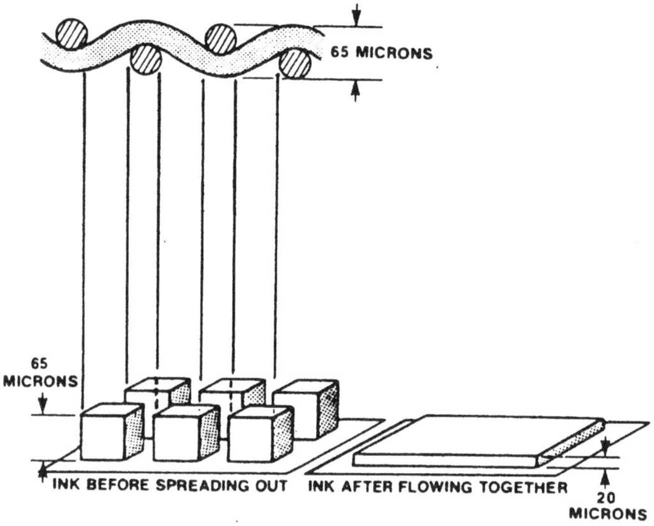

Fig. 11.17: Formation of film from ink dots

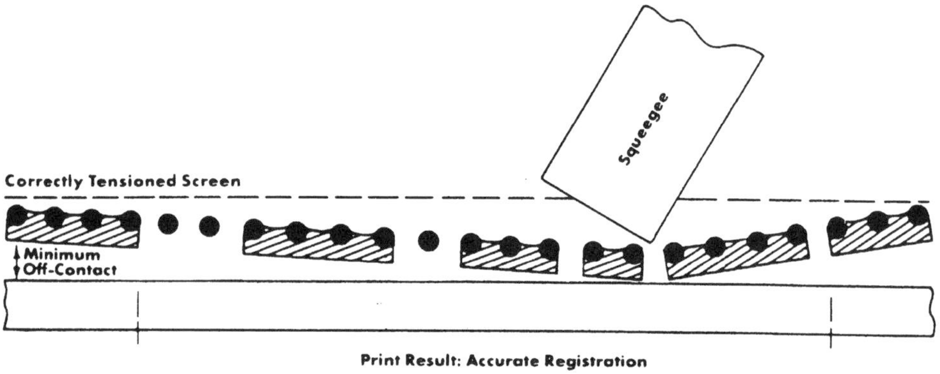

Fig. 11.18: Mechanics of screen-printing process. Courtesy of Tedko, Inc.

Screen printers range from basic hand-operated units to fully automatic machines. They are available in three styles: flat bed, rotary, and cylinder (Fig. 11.19). In the flat bed machine the printing stock and screen are stationary, the squeegee moves along the screen to deposit the ink. Production rates are up to 100 parts per minute.

The rotary screen printer has a drum that is constructed from a metal screen. The squeegee is positioned inside the drum and the ink is pumped into the inside. In this process the squeegee is stationary with both the screen and printing stock moving. The cylinder consists of a stationary squeegee and moving printing stock and screen. The process is similar to the rotary machine except stock is fed intermittently with rates up to 8000 pieces per hour.

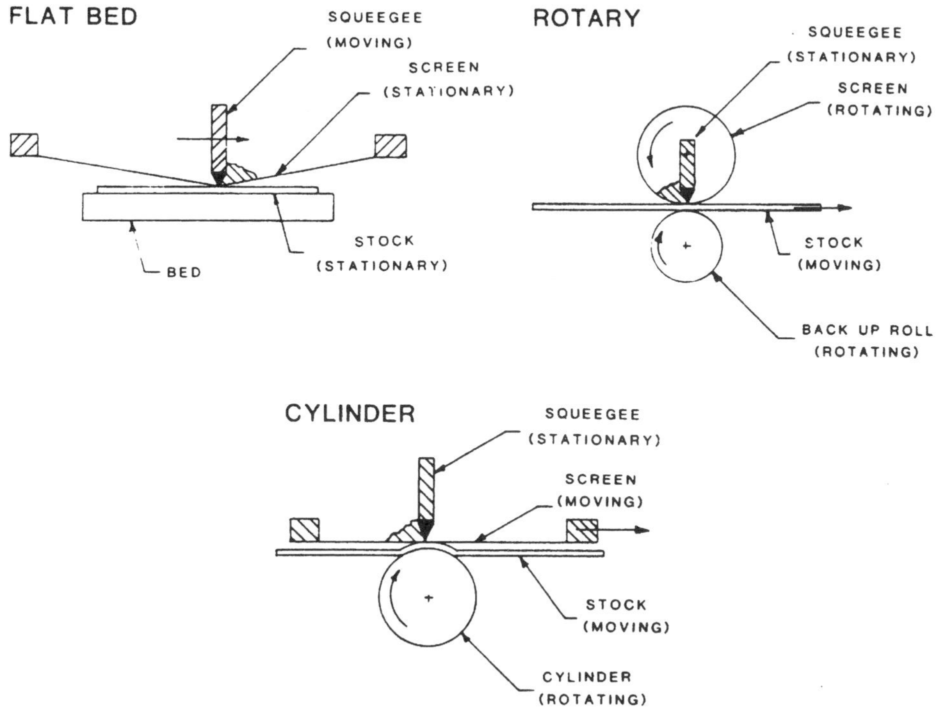

Fig. 11.19: Basic types of screen printers. Courtesy of Sheldahl, Inc.

11.13.6 Labels and Decals

Often manual application

UPC and bar codes have been responsible for the major part of the increased need for industrial blow molders to apply labels. Although other uses of labels and decals are required in industrial blow molding, quite often the most cost-effective method of application is by hand, due to irregular shapes and slow speeds, especially in comparison to packaging applications.

The use of pressure sensitive labels appears to be the overall most cost effective labeling method, and they are the easiest to use.

11.13.7 In-Mold Labeling

Cost effective for high-volume small containers

If the bottle or small part is under 6 liters (5 gallons), and the annual volume is 2 to 8 million units per year, then in-mold labeling (IML) should be considered. The initial investment is high and set up times much longer, but with the higher volume IML is competitive in price with post-mold labeling/decorative systems.

The basic IML process is suitable for rectangular and oval containers, and it usually employs paper labels produced from medium base stock coated on one side for printability and the other for heat seal coating. Die cut labels are placed in magazines from which robotic arms pick them up and insert them into the mold. Plastic labels have eliminated the bulge experienced with paper labels, and act as insulators and decrease the cooling rate of the plastic area covered by the label. A further advantage of plastic labels is that they are directly recyclable to the blow molding process. In-mold labels are printed by any one of the common printing methods.

Complete machine packages available

Setup of and maintaining IML systems has become easier in recent years because several machine manufacturers have developed molding machines that are better adapted to mounting labeling hardware, and they offer complete machine, molds, and labeling hardware packages. For example, Battenfield produces long stroke models that have more room for mounting IML equipment than the more commonly used swing arm machines or shuttle machines. Krupp Kautex also has a long stroke KBS-1 family of machines. Graham Engineering has introduced a novel type long stroke machine. The Graham Model 6000 twin has dual clamps side by side on a single carriage. Bekum has considerable success with IML in its B.M.-704 swing arm machine series (Fig. 11.20).

Molds that have been modified or designed for IML are fitted to both vacuum position and retain the "in-mold" labels. Owing to the negative air pressure effect (vacuum), such molds will usually have increased capacities proportional to the shrinkage reduction obtained resulting from the vacuum. When volumetric filling of such containers is employed, compensating adjustments in filling line height must be made. For example, a cubical container 152 mm (6 × 6 × 6 in) would require a 0.24 mm/1 cm (0.024 in/in) shrink allowance, and a blow mold cavity of 156 mm (6.144 × 6.144 in). If vacuum assist is exerted on the mold in conjunction with normal blow pressure of 60 to 90 psi (4.137 to 6.2055 bars) during blow molding the effect would be to produce a container having an outside finished dimension 154 mm (6.048 × 6.098 in) or an increase of 0.086 liter (5.23 in^3). Such an increase may result as a costly giveaway product (Fig. 11.21).

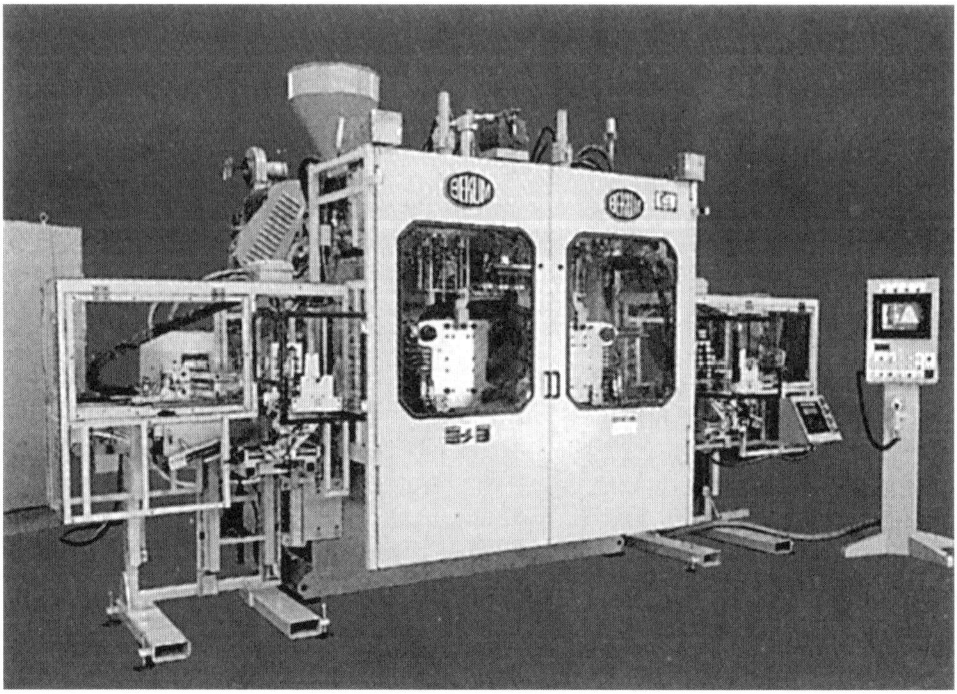

Fig. 11.20: In-mold label equipment

Fig. 11.21: In mold label operation

Additional Reading

McNamara, J., S.P.E. Blow Molding Workshop Manual. Fairfield, Connecticut. SPE 1975, ANTEC and SPE 1991, and Blow Molding RETEC 1991

Satas, D. (Ed.), *Plastic Finishing and Decoration.* (1986) Van Nostrand Reinhold, New York

Margolis, J. M. (Ed.), *Decorating Plastics.* (1986) Hanser, Munich

Stoeckhert, D. (Ed.), *Veredeln von Kunststoff-Oberflächen.* (1974) Hanser, Munich

Modern Plastic Encyclopedia. (1988) McGraw-Hill, New York

Stats, D., *Decorating Plastics.* In *SPI Engineering Handbook.* Berins, M. K. (Ed.) (1990) Van Nostrand Reinhold, New York

Van Hor, R.C., Spray Painting Plastics. In *Plastic Finishing and Decoration*, Satas, D. (Ed.) (1996) Van Nostrand Reinhold, New York

Knights, M., Is In-Mold Labeling in Your Cards? *Plastic Technology*, May 1995

Status, D., *Decoration Equipment and Processes*, In *Plastic Blow Molding Handbook*, Lee, N.C. (Ed.) (1990) Van Nostrand Reinhold

Rosato, D.V., Rosato D.V. (Ed.), *Blow Molding Handbook.* Hanser, Munich (1989)

12 Troubleshooting*

Troubleshooting is one of the most important functions in any molding operation. Each problem has its own solution, and each different type of machine, or even each machine of the same type, has its peculiarities. Recognize that no two machines operate in exactly the same manner and also that plastic materials do not melt as perfect blends, thus operating within permissible limits. The usual problem can be resolved with one or just a few changes during molding. Simplified guides to troubleshooting are given in Tables 12.1 and 12.2. More detailed processors' guides to troubleshooting are given in Tables 12.3, 12.4, and 12.5. Information on troubleshooting guides for granulators, conveying equipment, metering/proportional/feeding equipment, chillers, and dehumidifying dryer performance is given in Table 12.6.

A simplified approach to troubleshooting is to develop a check list that reviews basic rules for problem solving.

Check list

- Have a plan.
- Molding conditions should be watched.
- Change one condition at a time.
- Allow sufficient time at each change.
- Keep an accurate log of each change.
- Check housekeeping, storage areas, granulators, etc.
- Narrow the areas into which the problem belongs; that is, machine, dies/mold, operating conditions, material, part design, and management such as:
  - Change the material. If the problem remains the same, it probably is not the material.
  - Changing the type of material may pinpoint the problem.
  - If the trouble occurs at random, it is probably a function of the machine, the temperature control system, or the heating bands. Changing the die/mold from one press to the other permits a determination of whether it is in the machine, die/mold, and/or resin.
  - If the problem appears, disappears, or changes with the operator, observe the difference in their actions.
  - If the problem appears in about the same position of a single cavity mold, it is probably a function of the flow pattern and system.
  - If the problem appears in the same cavity or cavities of a multicavity mold, it is in the cavity.
  - If machine operation malfunctions, check hydraulic or electric circuits. As an example, a pump makes oil flow, but there must be resistance to flow to generate pressure. Determine where fluid is going. If actuators fail to move or move slowly, the fluid must be bypassing them or going somewhere else. Trace it by disconnecting lines if necessary. No flow (or less than normal flow) in the system will indicate a pump or pump drive is at fault. Machine instruction manuals will provide details concerning correcting malfunctions.

---

* *Source:* Rosato et al. (eds.), *Blow Molding Handbook*, 2nd ed., Carl Hanser Publishers, Munich (2004)

- Set up procedure to "break in" new mold:

Check list

  - Obtain samples and molding cycle information if mold is new to the shop but has been run before.
  - Clean mold.
  - Visually inspect mold. Obvious corrections, such as improving the polish or removing undercuts, should be done before the mold is put in service.
  - Check actions of the mold: try cams, slides, locks, unscrewing devices, and other devices on the bench.
  - Install safety devices.
  - Operate mold in press and move very slowly under low pressure.
  - Open mold and inspect it.
  - Dry cycle the mold without injecting material. Check knockout stroke, speeds, cushions, and low pressure closing.
  - After mold is at operating temperature, dry cycle again. Expansion or contraction of the mold parts may affect the fits.
  - Take a shot using maximum mold lubrication under conditions least likely to cause mold damage. These are usually low material feed and pressure.
  - Build up slowly to operating conditions. Run until stabilized, at least 1 to 2 hours.
  - Record operating information.
  - Take part to quality control for approval.
  - Make required changes.
  - Repeat process until approved by quality control and/or customer.

It is important to understand and to develop the proper approach in eliminating molding problems. The following review can assist the molder in finding the cause and probable remedy for problems occurring that result in unsatisfactory molded parts.

Basically, faulty or unacceptable molded parts usually result from problems in one or more of three areas of operation:

- Pre-molding: Material handling and storage
- Molding: Conditions in the processing cycle
- Post-molding: Parts handling and finishing operations

The solution to these problems is usually either quite obvious, or they are very specialized. This review discusses primarily the solution of problems encountered in the molding cycle. Causes of these faults can be grouped as follows:

- Machine
- Dies/molds
- Operating conditions (time, temperature, pressure)
- Material
- Part design
- Management

12.1 Detailed Failure Analysis

While millions of parts are produced by the blow molding process each year, it is not a simple processing technique. As a result, there are many types of defects that can appear in molded parts. Although this chapter addresses these problems as they relate to lightweight, thinwalled products such as containers, many of the recommended solutions presented here also apply to heavier, thicker parts. This section provides information on some typical conditions for evaluation.

Overview and Summary

Troubleshooting PE bottles

Table 12.1: Simplified Guide to Troubleshooting Polyethylene Bottles

Difficulty	Suggested solutions
Low gloss	• Increase material temperature
	• Increase mold temperature
	• Improve mold venting
	• Increase air pressure
	• Increase air blowing rate
Excessive cycle	• Decrease material temperature
	• Decrease mold temperature
	• Decrease part wall thickness
	• Improve mold coring
	• Increase material density
	• Increase air pressure
	• Increase melt index (with decreased stock temperature)
Die lines	• Clean the die
	• Smooth and polish the die
	• Increase mold temperature
	• Increase air pressure
	• Increase air blowing rate
	• Increase purge time when changing materials
	• Improve die streamlining
Low bottle weight	• Increase die to mandrel clearance
	• Decrease material temperature
	• Increase extrusion speed
	• Decrease extrusion die temperature
Surface roughness	• Increase stock temperature
	• Decrease extrusion speed
	• Increase die temperature
	• Improve die streamlining

(continued)

Table 12.1: *(continued)*

Difficulty	Suggested solutions
Weak pinch	• Adjust material temperature
	• Increase pinch blade land width
	• Decrease rate of mold closing
Parison curl	• Adjust and center die-parison curls toward thin area
	• Improve die and head heat uniformity
Wall thickness non-uniform	• Decrease material temperature
(a) Vertical (sag)	• Increase extrusion speed or accumulator ram pressure
	• Lower melt index
	• Increase material density
(b) Circumferential	• Decrease blowup ratio
	• Improve head and die heat uniformity
	• Adjust and center the die
Excessive thinning at parting line	• Decrease stock temperature
	• Increase mold temperature
	• Increase air pressure
	• Increase air blowing rate
	• Improve mold venting
	• Improve mold temperature
Excessive shrinkage	• Decrease wall thickness, or make more uniform
	• Decrease mold temperature
	• Increase air pressure
	• Decrease material temperature
	• Decrease density
	• Improve mold coring
Excessive parison swell	• Increase material temperature
	• Increase die temperature
	• Decrease extrusion speed
	• Increase melt index
	• Reduce die size
Doughnut formation	• Wait for mandrel temperature to reach die temperature
	• Clean the lower surface of the die
Warped top and bottom	• Slow the cycle
	• Decrease mold temperature
	• Decrease stock temperature
	• Decrease part weight
	• Improve mold coring

(continued)

Table 12.1: *(continued)*

Difficulty	Suggested solutions
Variable bottle weight	• Finer extruder screen pack • Increase screw cooling • Raise rear extruder heats • Decrease extrusion rate

Troubleshooting PE dairy gallon

Table 12.2: Simplified Guide to Troubleshooting PE Dairy Gallon Based on Parison, Structural, and Appearance Problems

(a)

Parison problems	**Causes**
Die lines or streaking	• Foreign matter lodged in die gap
	• Burned-out heater band
Parison pleating or curtaining	• High melt temperature
	• High weight-swell resin
	• Misaligned die
	• Short land between mandrel and bushing
Parison curl	• Cold bushing or mandrel
	• Misaligned die
	• Foreign matter on die
Parison stringing	• High melt temperature

(b)

Structural problems	**Causes**
Thin, weak welds	• High melt temperature
	• High mold temperature
	• Short or sharp pinch-off land
Cutting at pinch-off	• Low melt temperature
	• Fast mold close
Structural problems	**Causes**
Blowouts	• High melt temperature
	• High pinch-off temperature
	• Insufficient cooling
	• Parison stringing
	• Low clamp pressure
	• Short parison
	• High blow-up ratio
	• Low diameter-swell resin
Webbing in neck or upper handle	• Parison pleating
Warpage	• High melt temperature
	• Short cooling cycle
	• Blocked cooling channels
	• Incorrect parison programming
Tearing of flash	• Long pinch-off land
	• Low clamp pressure

(continued)

Table 12.2: *(continued)*

(c)

Appearance problems	Causes
Foreign matter in container wall	• Contamination resin feed
	• Poor melt path streamlining
	• Mismatched head sections
	• Poor fit between die/adaptor/extruder
Bubbles in container wall	• Low melt temperature
	• Moisture in resin
	• Mismatched head sections
	• Low back pressure
Sharkskin container surface	• Wrong parison drop rate
Cold spots in container wall	• Insufficient working of melt
	• Low back pressure
	• Low melt index resin mixed into feed
Poor mold surface reproduction	• Low melt temperature
	• Low blow pressure
	• Smooth mold surface
	• Inadequate venting at mold parting line
	• Air leak around blow pin
Pitting of container surface	• Moisture in resin
	• Moisture on mold surface

Part defects

Table 12.3: Troubleshooting Guide for Extrusion Blow Molding. (a) Part Defects

Cause	Possible remedy
Blowouts	
• Pinch-off too sharp which cuts parison and causes blow-out	• Increase width of pinch-off land
• Pinch-off too wide	• Decrease width of pinch-off land
• Clamp pressure too low	• Adjust clamp pressure
• Pinch-off too hot	• Check cooling water-flow to mold • Check stock temperatures • Check cooling channels for blockage • Check cycle time • Check part design
• Blow pressure too high	• Gradually reduce pressure to avoid stretching the parison at a rate too high
• Parison too short	• Ensure proper parison length
• Blow-up ratio too high	• Check head tooling (change to a larger diameter die may be required)
• Resin swell too low	• Use higher swell resin to ensure proper parison size and to prevent missed handle, accompanied by blow-out
• Moisture in resin	• Check for moisture in resin
• Fines in hopper	• Eliminate hopper fines
• Stock temperature too low	• Increase stock temperature
• Extruder feed zone temperature too low	• Increase extruder feed zone temperature
• Blowing air pressure too high	• Adjust blowing air pressure
Bubbles	
• Moisture in melt (because of condensation appearing on cold resin exposed to warm atmosphere)	• Properly dry hygroscopic resin • Permit resin to warm and evaporate before entering extruder
• Moisture in melt caused by too much water cooling on the throat, producing condensation in the barrel and moisture in melt when running at slow screw speeds	• Adjust throat cooling
• Air inclusion mismatched head sections	• Align head sections and/or increase back pressure
• Contamination	• Check and eliminate contamination source
Cap leakage	
• Worn or damaged blow pin or shear steel	• Replace blow pin and/or shear steel
• Blow pin alignment	• Align blow pin
• Mold alignment (mismatch)	• Replace guide pins and bushings

(continued)

Table 12.3: *(continued)*

Cause	Possible remedy
• Wrong blow pin and shear steel	• Replace with correct blow pin and shear dimensions steel
• Wrong neck plate dimensions	• Replace neck plates
• Wrong cap dimensions	• Check and ensure correct size caps are being used
• Pleating (vertical web inside neck)	• Adjust preblow system
• Capper tongue too high/too low	• Adjust capper tongue
Cold spots or marbleizing	
• Insufficient back pressure	• (Reciprocating unit) Adjust back pressure gauge
	• (Continuous extrusion unit) Change screen pack to finer mesh
• Mixture of resin having a lower melt index	• Avoid improper resin mixing
• Melt too cold	• Increase melt temperature
• Cycle too fast	• Increase blow time
Container does not blow or incomplete blowing	
• Blow air restricted	• Check air system
• Pinch-off area too sharp	• Provide broader pinch-off area
Dirty or streaked pieces	
• Contamination	• Inspect storage and heating areas of cleanliness
	• Purge die and clean hopper
Doughnut forming between top of mold and die face during blowing	
• Back pressure on screw set too high	• Slowly decrease screw back pressure
• Die tip temperature too high	• Reduce die tip temperature
• Melt temperature too high	• Slowly decrease melt temperature throughout system
Excessive flash	
• Parison diameter too large	• Change tooling
• Flash pockets too shallow	• Consult mold maker
• Improper mold closure	• Increase locking and check for mold mismatch
Excessive shrinkage	
• Blown part ejected when too hot	• Reduce mold temperature
	• Reduce melt temperature
	• Increase blow time
• Inadequate mold cooling	• Clean mold cooling channels

(continued)

Table 12.3: *(continued)*

Cause	Possible remedy
Hole or slit in parting line	
• Melt temperature too low	• Increase stock temperature
• Mold closing too rapidly	• Cushion model closing speed
• Maladjusted trimming equipment	• Adjust trimming equipment
• Excessive pinch-off land length	• Reduce pinch-off land length
Improper part weight	
• Mandrel in up or down position	• Adjust mandrel to increase or decrease part weight
• Continuous extrusion, rpm is too high or too low	• Increase or reduce extruder rpm; note: using a longer die/mandrel also may increase part weight
• Material or tooling too hot	• Reduce stock temperature
	• Reduce head temperature
• Rupturing blow air	• Check blow pin and shear steel
• Stripper not properly adjusted	• Adjust stripper plate
• Blow pins set too low or too high	• Raise or lower blow pins
Irregular striations and other marks	
• Tube chafing on die	• Align parison to die
Lost handles	
• Misaligned molds	• Realign molds to catch handles
• Parison hooking	• Realign die adjusting ring
• Parison melt stringing	• Reduce melt temperature
• Low flare	• Lower stock temperature
	• Increase shot pressure
Low environmental stress crack resistance	
• Excessive percentage of regrind	• Reduce regrind rate
• Melt (stock) temperature too low	• Increase melt (stock) temperature
• Melt (stock) temperature too high	• Decrease melt (stock) temperature
• Mismatched molds	• Realign molds
• Worn or damaged trimmer nests	• Replace trimmer nests
• Incorrect resin for application	• Use proper resin
Low impact strength	
• Excessive percentage of regrind	• Reduce regrind rate
• Mismatched molds	• Realign molds
• Worn or damaged trimmer nest	• Replace trimmer nests

(continued)

Table 12.3: *(continued)*

Cause	**Possible remedy**
Low top load strength	
• Uneven wall distribution	• Reduce parison drop time
	• Reprogram parison programmer
• Low part weight	• Increase part weight
• Incorrect resin for application	• Use proper resin
Nonuniform wall section circumference	
• Inaccurate die centering	• Adjust die to provide uniform distribution of wall section
• Unsuitable blow head	• Check blow head design
• Head temperature not uniform	• Stagger heater band gaps on head
• Loose die pin	• Tighten die pin
Part too hot	
• Mold temperature too high	• Decrease mold temperature
• Cycle time too short	• Increase blow time to extend cycle time
• Clogged coolant lines/channels	• Clean lines/channels
Part volume too low	
• Part weight too high	• Reduce part weight
• Mold too hot	• Reduce mold temperature
• Cycle time too short	• Increase cycle time
• Mold volume incorrect	• Resize mold
• Volume-reducing inserts used	• Remove inserts for packed parts only
• Blow air pressure too low	• Increase blow air pressure
• Part storage areas too hot	• Reduce storage area temperature or dwell time
• Melt temperature too high	• Reduce melt temperature
Part volume too high	
• Part weight too low	• Increase part weight
• Mold too cold	• Increase mold temperature
• Cycle too long	• Reduce cycle time
• Wrong mold volume	• Resize mold
• Volume-reduction inserts not used	• Install inserts
• Blow air pressure too high	• Reduce blow air pressure
• Part storage area too cold	• Increase storage area temperature or dwell time

(continued)

Table 12.3: *(continued)*

Cause	Possible remedy
Poor definition of detail	
• Blow air pressure too low	• Increase blow air pressure
• Poor mold venting	• Consult mold maker
Poor part surface (roughness, pits, orange peel, etc.)	
• Poor mold surface	• Check mold surface (It should have a fine matte finish to permit air to vent quickly and parison to conform to mold surface while still hot enough to do so)
• Inadequate vents	• Check vents along mold parting line; escape of air minor reworking may be necessary
• Low blow pressure or blow rate	• Increase blow pressure
	• Ensure that blow pin is large enough to handle required amount of air to fully and rapidly blow part
	• Check for restrictions or partial plugging of air lines
• Air leak around blow pin	• Check for leakage which prevents sufficient pressure to hold part against the mold for conformation and rapid cooling
	• Check blow pin and/or shear ring and replace if necessary
• Low stock temperature	• Gradually increase stock temperature but avoid overly high stock temperature
• Condensed water in mold	• Check for low mold temperature, combined with environmental humidity (If mold temperature cannot be raised, investigate plant humidity control)
	• Reduce cooling water flow
• Too much mold lubricant	• Use less lubricant
• Melt temperatures too low	• Increase melt tempreature
	• (Extruder) if surface roughness appears, which may be indicative of melt instability, reduce extrusion pressure until problem disappears
	• (Reciprocating or accumulator) drop out of melt instability region by further decreasing extrusion rates (Note: Generally, an unsatisfactory solution because of productivity decrease)
	• Increase extrusion rates (decrease drop time)
	• Increase stock temperature or change to higher melt index resin

(continued)

Table 12.3: *(continued)*

Cause	Possible remedy
Poor weld or pinch-off	
• Thinning of weld	• Check stock and mold temperatures
• Tearing of flash during trimming	• Check pinch-off and length (Recommended land lengths of 0.010 in to 0.015 in are common for polyethylene.)
• Cutting at pinch-off	• After checking weld and pinch-off, check for hole or slit along weld which may indicate that stock temperature is too low, thus causing parison to tear as mold closes
• Seams not uniform because of improper line up of mold and parison	• Ensure that blow head is located exactly vertical and improve adjustment of mold to die
	• Meter tube faster
	• Adjust mold closure time
	• Increase stock temperature
• Parison temperature too high	• Decrease parison temperature
• Pinch-off area too hot	• Decrease mold temperature at pinch-off area
Short molding	
• Insufficient hold time on blowing pressure	• Increase blowing time
Sticking in mold	
• Mold temperature too high or premature opening of mold	• Improve coring and heat transfer and slightly reduce mold temperature
	• Increase cooling time
• Parison hooking	• Adjust parison drop
• Swing arm out of adjustment	• Adjust swing arm
• Molds not opening far enough	• Adjust mold open stop position
• Tail too short	• Increase screw rpm
	• Increase blow time
• Mold too hot	• Check mold cooling system
• Insufficient exhaust time	• Increase exhaust time
Tail lengths not uniform	
• Manifold chokes out of adjustment	• Adjust manifold chokes making short tails longer first
• Part weights not uniform	• Adjust part weight to ±?g of target weight
• Non uniform head heats	• Check heaters
• Head chokes out of ajustment	• Reset heat chokes

(continued)

Table 12.3: *(continued)*

Cause	Possible remedy
Thin parting line	
• Mold not fully closed	• Clean mold faces • Adjust clamp pressure • Set downward travel of blowpins • Decrease blow air pressure
Tail flash not separating from part	
• Parison too short	• Lengthen parison
• Flash pocket too deep	• Contact mold maker
Top flash not separating from part	
• Dull cutting ring	• Sharpen or replace cutting ring
• Insufficient contact between cutting ring and striker plate	• Adjust down position of blow pin • Increase downward force of blow pin
Unmelted pellets in blown part	
• Melt not homogeneous (insufficient mixing in screw)	• Increase back pressure by choke adjustment • Use finer screw pack
• Melt temperature too low	• Increase heater setting
Vertical lines	
• Foreign matter or degraded resin in head	• Clean die gap • Open to largest die gap and purge • Remove and clean tooling
• Damaged tooling	• Repair or replace tooling as needed
Wall section too thin at die orifice end	
• Excessive draw-down because extrusion rate is too low, melt temperature is too high, or expansion is too late	• Increase extrusion rate • Decrease melt temperature, close mold and expand more rapidly
Wall section too thin in general	
• Improper parison size	• Enlarge parison • Reduce melt temperature • Meter faster

(continued)

Table 12.3: *(continued)*

Cause	Possible remedy
Warpage (rocker bottoms and oval necks)	
• Insufficient mold cooling	• Determine whether an increased water flow will stop warpage
	• Reduce melt temperature
	• Increase cycle time to provide longer cooling
• Stock temperature too high	• Drop temperature by small increments to see if it will affect a cure without causing new problems such as cold spots in container wall
• Blocked cooling channels	• Check throughput and clean channels if throughput is significantly less than when mold was new
• Cycle time too short	• Adjust cycle time as needed
• Poorly designed part	• Check material distribution in part for unnecessarily thick or thin sections
Webbing	
• Incorrect mold position relative to parison	• Change position of mold relative to parison
• Incorrect parison diameter	• Change tooling
• Internal parison walls touching	• While mold is closing, add prepinch bars to mold

Parison Defects

Table 12.3: Troubleshooting Guide for Extrusion Blow Molding (b) Parison Defects

Cause	Possible remedy
Blowout of parison	
• Pinch-off too sharp or hot	• Increase land width of pinch-off
	• Increase cooling, preferably in pinch-off area
• Blow pressure too high, parison inflation rate too rapid	• Reduce blow pressure and/or inflation rate
• Insufficient mold clamping	• Increase clamp pressure
• Parison too short and not fully captured in pinch-off	• Increase shot
	• Close mold before full snapback occurs
• Blow-up ratio too high	• Check die head size
Bubbles	
• Small bubbles may be caused by moisture in resin	• Follow correct drying procedures for hygroscopic resins
	• Make provisions for keeping pellets hot in hopper to prevent pick-up of moisture

(continued)

Table 12.3: *(continued)*

Cause	Possible remedy
• Large bubbles which sometimes burst while leaving die orifice may be caused by trapped air	• Increase back pressure • Raise screw speed if possible
Curtaining and webbing	
• Poor tooling design	• Check land length between bushing and mandrel
• Die misalignment	• Check bushing and mandrel alignment to ensure that different flow rates in different areas are not occurring
• High stock temperature	• Avoid high stock temperatures which will yield a parison that will collapse before it can be blown
• Resin mismatch	• Avoid choosing resin with a weight swell too high
• Parison wall too thin	• Increase die gap
Die lines or streaking in parison	
• Contaminated or degraded resin	• Check for dirt, dust and lint in unprocessed resin • If using regrind, check for foreign matter
• Burned-out heater band (Evidence will appear as rippled streaking in parison causing cold spots)	• Replace non-functioning band
• Foreign matter lodged between die and mandrel	• Increase die gap briefly to purge particles, then reset to original gap
• Carbon or degraded resin in head	• Clean die and mandrel with copper, brass, or material softer than machine parts to prevent marring and scratching
• Melt overheating	• Check blow head, accumulator, and cylinder for dead areas and pockets • Check mandrel for slight gaps with contact surfaces causing hang-up • Check for cylinder hot spots and overheating and reduce temperature or eliminate hot spot
• Damaged tooling	• Repair or replace tooling as needed
Dull, rough, or striated melt surface	
• Insufficient melt or stock temperature	• Increase die temperature and stock temperature to produce melt which is clear and glossy when leaving die orifice
• Incorrect extrusion rate	• Adjust rate
• Dirty die/head tooling	• Clean tooling
• Melt fracture	• Increase melt temperature • Decrease machine rpm • Increase die gap

(continued)

Table 12.3: *(continued)*

Cause	Possible remedy
Excessive tubular melt stretching after emerging from die orifice	
• Melt temperature too high	• Reduce die temperature and/or stock temperature
• Melt extrusion too slow	• Increase extrusion rate
Hole in parison	
• Contamination	• Check for degraded material • Check for hang-up in tooling, head, or screw • Check for foreign material in resin
• Trapped air	• Permit extruder to run for a few minutes; problem could be caused by letting hopper run out of resin before refilling
• Moisture in resin	• Properly dry hygroscopic resin before using
Horizontal rings	
• Programming changes too drastic	• Decrease flow to programming cylinder • Decrease weight change between steps
Melt fracture (sharkskin)	
• Melt flow is unstable	• Increase or decrease stock temperature • Increase or decrease drop time • Check to ensure that head choke and manifold chokes are open as much as possible • Clean tooling
Nonuniform parison	
• Improperly designed blow head	• Check blow head design
• Nonuniform heating in blow head	• Check heat distribution in blow head
• Mandrel improperly centered in die ring	• Center die
• Melt flow too rapid	• Reduce extrusion rate
• Melt temperature too high	• Reduce stock temperature
• Die land length too short	• Use longer land die mandrel
Parison curl and doughnutting	
• Cold mandrel or bushing	• Permit sufficient warm up time before attempting production • Ensure that die bushing heater is operative, if so equipped
• Misaligned die and mandrel	• Check positioning of die and mandrel; die bushing may need to be machined to bring mandrel to a flush or lower position with die face

(continued)

Table 12.3: *(continued)*

Cause	Possible remedy
• Foreign matter or degraded resin in die bushing	• Clean die
• Damaged tooling	• Check tooling for damage
• Improper knife cut	• Increase distance between knife and die face
	• Check knife edge for sharpness and cleanliness
Parison deflects without becoming punctured	
• Needle is blunt	• Sharpen needle or increase penetration speed
Parison hooking	
• Nonuniform parison walls	• Center adjusting ring around mandrel to correct
• Die mandrel channel dirty	• Clean channel
• Defective head/die heaters	• Replace heater
• Loose die pin	• Tighten die pin
• Head temperature not uniform	• Stagger heat band gaps on head
• Die gap out of adjustment	• Adjust die around mandrel
• Air flowing on parison	• Shield parisons from moving air
• Warped die or mandrel	• Replace damaged component
Parison sag	
• Melt temperature too high	• Decrease melt temperature
• Resin melt index too high	• Use lower melt index resin
• Extrusion rate too slow	• Increase extrusion rate
• Parison drop time too long	• Decrease parison drop/hang time
• Parison weight too heavy	• Reduce parison weight
• Mold closing too slowly	• Increase low pressure close
	• Lubricate tie bar bushing
Parison stringing	
• Stock temperature too high	• Gradually lower stock temperature
• High fill pressure	• Reduce fill pressure until weeping stops
Pinch-off too thick	
• Pinch-off edges too wide or inadequate mold closure	• Sharpen pinch-off edges
	• Increase mold closing pressure
	• Check mold parting line for poor match
Rolling	
• Melt clings to die or mandrel and flows to coolest side	• Adjust die temperature

(continued)

Table 12.3: *(continued)*

Cause	Possible remedy
Smoking	
• Heat controller malfunction	• Check heat controllers
• Contamination in resin	• Check for contamination
• Stock temperature too high	• Decrease stock resin temperature
• Extrusion pressure rate too high	• Decrease extrusion pressure rate
• Ratio of regrind to virgin material too high	• Decrease regrind level

Table 12.3: Troubleshooting Guide for Extrusion Blow Molding (c) Complete Process Summary

Problem	Condition	Cause	Solution
Heating interrupted	Temperature falling	(a) Control action to avoid overshoot, timeproportioning, or interaction with adjacent zones	Observe if heating resumed in 30 s, otherwise check (b)
		(b) Fail-safe control, indicating broken thermocouple or connection	Check thermocouple continuity
Temperature continues to rise	Temperature above upper control limit	(a) Thermocouple slipped out or not in contact with heater/extruder	Replace, ensuring good thermal contact
		(b) Thermopouple incorrectly connected	Check connections and temperature indication, e.g., in hot water or body heat
Output suddenly falls	Steady screw speed	(a) Feed hopper empty	See barrel and feed unit section in this chapter; investigate blocked feed line
		(b) Bridging in hopper or into screw	Stop screw and clear with non-metallic rod, e.g., strip of product; check adequate feed pocket cooling; check air entrapment esp. with powders
		(c) Polymer melting in feed section	Reduce first barrel zone temperature Increase feed pocket cooling Cool feed end (only) of screw
		(d) Surging	

(continued)

Table 12.3: *(continued)*

Problem	Condition	Cause	Solution
Polymer leakage from joints or fittings	Excessive pressure	(a) Blocked die or screen	Immediately reduce screw speed Check die and adaptor temperature Shut down and clear
		(b) Failed die heater or control	Immediately reduce screw speed Check thermocouple Shut down and replace heater
		(c) Incompletely melted polymer, esp. at start-up or after feed interruption	Stop screw Wait to heat up and restart *slowly*
Screw stops or speed drops	Excessive torque	(a) Low barrel temperature	Reduce speed setting and check heaters
		(b) Excessive pressure	As for "Polymer leakage" above
		(c) Excess feed, e.g., of strips	Stop or reduce speed setting Increase temperature of first barrel zones
		(d) Excessive speed	Reduce speed and/or increase barrel temperatures
Lumpy or unmelted polymer at die	Incomplete melting, esp. of powders	(a) Insufficient shearing/ heating in barrel	
		(b) Excessive screw speed	Reduce screw speed *or* improve melting
		(c) Feed contamination, e.g., with hard-grade polymer	Reduce screw speed Check feed material if necessary, increase barrel temperatures
Foaming or bubbles at die (distinguish from later contraction bubbles due to rapid cooling), surface blisters	Steady running or start-up	(a) Moist polymer	Stop feed and restart with dried polymer
		(b) Excess temperature causing decomposition	Stop or reduce screw speed
		(c) Air entrapment in feed	Raise first barrel zone temperature (to reduce unmolten length) Fit vent at rear of feed opening

Table 12.4: Troubleshooting Guide for Injection Blow Molding

Cause	Possible remedy
Cocked necks	• Movable bottom plug stuck in • Reset stripper plate • Reduce inject portion of cycle • Increase blow air time/pressure
Color streaks-flow lines	• Raise back pressure • Increase injection pressure • Increase melt temperature • Open nozzle orifice • Change color mix • Change color batch • Add mixing pin to screw • Reduce injection speed • Increase parison mold temperature • Dry material
Contamination (oil/grease on part)	• Wash parison and blow molds with solvent (especially in neck rings) • Wash core rods • Clean air filter • Replace O-rings between molds
Cracked necks	• Increase melt temperature • Increase neck temperature in parison mold • Reduce core rod cooling • Increase neck temperature in blow mold • Reduce retainer grooves in core rods • Movable bottom plugs stuck in (regrind materials only) • Increase injection speed • Balance nozzles for even fill • Check core rod alignment • Check operation of mold temperature controller • Check stripper location and speed (especially in stryrene) • Open nozzle orifice • O-rings in face blocks

(continued)

Table 12.4: *(continued)*

Cause	Possible remedy
Dimensional problems	
• H – height	• Increase by moving parison and/or blow mold out (add shim)
	• Reduce by moving parison and/or blow mold in (remove shim)
• S – neck finish	• Increase by moving parison mold out (add shim)
	• Reduce by moving parison and/or blow mold in (remove shim)
• T – average of two dimensions	• Lower parison mold neck temperature
• To raise T	• Increase injection time
	• Increase stabilize time
	• Lower blow mold neck temperature
	• Increase core rod cooling (internal)
• To lower T	• Increase parison mold neck temperature
	• Increase blow mold neck temperature
	• Reduce core rod cooling
	• Hot line blow mold neck (Note: in some cases the opposite will happen when above is done when the T is being blown out in the blow mold)
• E – taken across major and minor axes	• Refer to T dimension
Distorted shoulder	• Increase blow air pressure
	• Shorten cycle
	• Wash out vents in blow mold
	• Increase parison mold temperature
	• Clean or replace (plugged) core rods
	• Increase blow air time
	• Reset or replace stripper
Engraving	• Increase temperature of body mold temperature controller
	• Increase melt temperature
	• Sandblast molds
	• Adjust engraving depth and width
	• Adjust vents
	• Increase blow air pressure
	• Clean out engraving
	• Increase blow delay
	• Balance nozzles for even fill

(continued)

Table 12.4: *(continued)*

Cause	Possible remedy
	• Clean, check, and set all gaps evenly in core rod valve
	• Check mold temperature controller operation
	• Increase venting on blow mold
	• Adjust parison mold temperature
	• Adjust blow mold temperature
	• Increase core pin cooling
Heavy in center	• Increase temperature of gate mold temperature controller (parison mold)
	• Move nozzles in
	• Decrease core rod cooling air
Hot spots	• Increase core rod cooling
	• Reduce injection pressure
	• Reduce melt temperature
	• Lower parison mold temperature in affected area
	• Reset stripper for localized external cooling of core rod
	• Check mold temperature controller operation
Inconsistent shot size	• Increase back pressure
	• Check hydraulic injection pressures for variations
	• Increase screw recovery time
	• Balance nozzles
	• Check for loose or bad thermocouples
	• Check for broken element in mixing nozzle
	• Check mold temperature controller operation
	• Adjust screw rpm
Nicks	• Replace damaged blow mold
	• Clean plastic or dirt out of blow mold cavity
	• Repair or replace damaged parison mold
	• Replace or repair damaged core rod
	• Remove strings from nozzles
	• Remove burrs from parison and blow molds
	• Repair or replace damaged parison and blow molds
	• Wash out vents in blow mold
	• Sandblast blow mold
	• Reset mold in die set

(continued)

Table 12.4: *(continued)*

Cause	Possible remedy
Nozzle freeze-off	• Remove contaminated material from nozzle • Raise temperature of gate mold temperature controller • Increase manifold temperature • Increase melt temperature • Reduce cycle • Open nozzle orifice • Check manifold heaters, fuses and wiring
Ovality of T & E dimensions (uneven shrinking)	• Increase temperature in shoulder/body area of parison mold • Reduce temperature in neck of parison mold • Lower blow mold neck temperature • Increase melt temperature • Check operation of temperature control unit
Push-up depth	• Increase blow time/pressure • Adjust mold/bottom plug cooling • Lower melt temperature • Clean and check core rod valve; set all gaps evenly • Check movable plug if used • Increase core rod cooling • Decrease air pressure • Adjust nozzle • Increase tip cooling
Saddle finish (usually apparent with oval T and E – try to correct oval T & E)	• Parison not packed up tight enough; add inject or screw time and/or pressure • Increase retainer grooves in core rods • Reduce parison mold neck temperature • Increase or decrease cushion • Increase holding pressure • Increase cooling time
Short shots	• Clean nozzle • Open nozzle orifice • Increase high injection pressure • Increase packing pressure • Increase injection time • Increase melt temperature

(continued)

Table 12.4: *(continued)*

Cause	Possible remedy
	• Raise all parison mold temperatures • Increase back pressure • Increase screw recovery stroke • Lengthen cycle time • Check and clean non-return valve on end of extruder screw • Adjust screw rpm • Increase screw speed • Increase cushion • Clean out hopper and throat
Sticking of parison to core rods (core rod too hot)	• Decrease cushion • Reduce melt temperature • Pack parison harder (increase injection pressure/time) • Reduce screw speed • Add stearate (release agent) • Increase back pressure • Adjust core rod temperature • Check mold temperature controller operation • Check for folds in bottom • Increase internal and external air cooling
Stripping difficulties	• Increase stripping pressure • Check core pins for burrs • Add lubricant to polymer • Adjust stripper bar alignment • V groove too deep
Sunken panels	• Increase blow air time • Reduce blow mold temperature • Shorten inject/transfer portion of cycle • Reduce core rod cooling • Add sink correction to blow mold • Check mold temperature controller operation
Thin parts	• Lower temperature of gate mold temperature controller in parison mold • Check core rod lock-off in parison mold • Check nozzle seats

(continued)

Table 12.4: *(continued)*

Cause	Possible remedy
	• Reset nozzles • Check parison mold part line • Lower melt temperature • Move nozzles out • Add injection and/or screw time • Replace nozzles • Check to see that mold temperature controller is functioning properly
White or black marks on neck finish caused by gas burning	• Lower melt temperature • Reduce injection speed • Lower temperature in neck of parison mold • Vent (relief) neck ring of parison mold • Open nozzle orifice • Check mold temperature controller operation • Reduce ram speed

Troubleshooting for stretched injection blow molding

Table 12.5: Troubleshooting Guide for Stretched Injection Blow Molding

Cause	Possible remedy
Air bubbles in the preform	
• Air entrapment due to too much decompression in plastifier	• Check dryer settings • Increase back pressure slightly
Bands of thick and thin sections in part wall	
• Improper settings on heat zones (zones colder than others)	• Ensure uniform temperature from capping ring to tip
• Not enough time for equilibration of preform before blowing	• Equilibration time • Change heat zones
Bands of vertical strips on the preform	
• Too much heat at specific area on preform	• Reduce heat
• Improper rotation for vertical bands	• Check rotation speed
Blemishes on part	
• Dirt in molds	• Blow molds should be cleaned
• Water droplets forming in molds or sweating causing condensation	• Increase mold temperature slightly to alleviate condensation
Cloudiness	
• Melt temperature too low	• Increase melt temperature slightly
• Moisture in injection molds	• Check for condensation on cores or in cavities • Increase water temperature in injection molds
Drag marks on preform	
• Injection mold damaged or scratched	• Polish and possibly rechrome cores and cavities
Fish eyes or zippers	
• Scratch marks on preform surface (normally caused by preforms contacting each other after being ejected from injection mold while still hot)	• Minimize contact of preforms after injection and prior to cooling
Folds in neck area	
• Center rod stretching preform too early	• Synchronize center rod stretch with air delay timers to get blow air to enter at correct time
Heavy material in bottom of part	
• Improper preform design	Redesign preform
• Improper cooling in molds causing heavy amount of materials to shrink back	• Improve mold cooling

(continued)

Table 12.5: *(continued)*

Cause	Possible remedy
Knit lines appearing in preform	
• Melt not being injected fast enough or hot enough	• Increase temperature • Raise injection pressure
Long gates	
• Valve gates in mold operating improperly	• Clean mold
• Incorrect temperature in hot runner system	• Check thermocouples
Mismatch lines on the preform	
• Mold misalignment	• Check cores and cavities for alignment
Off-centered gates	
• Could be caused by not using center rods	• Use center rods
• Poor concentricity in the preform (preform concentricity should be held to 0.005 in. max.)	• Check injection mold for concentricity of core rod and cavity • Reduce injection speed and injection pressure
Pearlescense (haze in container)	
• Preform stretching too fast for heat in preform	• Increase heat in area showing pearlescense going back to preform
Preform drooling	
• Valve gates are too warm	• Decrease temperature to valve gates
• Not enough packing pressure or time	• Increase hold time
Radial rings on preform	
• Moisture condensation on core rods	• Increase water temperature in molds
Scratches on part	
• Possible drag marks on preforms from cavity or core of preform	• Polish core and cavity of preform mold
Soft necks or deformed capping rings on finished container	
• Too much heat in top area of preform	• Reduce heating in affected zone. Heat shield may be added to shield capping rings
Undersized parts	
• Not enough high pressure air blow time (Container not being blown to side wall and held under high pressure to freeze material and set outline of mold)	• Check mold cooling • Check blow pressures and time • Check vents on mold
Yellowing of preform (indicating oxidization through excessive heating during drying)	
• Check drying temperature and time	• Adjust drying time as required

Troubleshooting auxiliary equipment

Table 12.6: Troubleshooting Guide for Auxiliary Equipment

(a) Dryers

Problem	Possible causes	Suggested solutions
Process air temperature too low	• Incorrect temperature selected on control panel	• Dial in correct temperature
	• Controller malfunction	• Check electrical connections, replace controller if necessary
	• Process heating elements	• Check electrical connections, replace elements if necessary
	• Hose connections at wrong location	• Check to make sure delivery hose is entering bottom of hopper
	• Supply voltage different from dryer voltage	• Check supply against nameplate voltage
Process air temperature too high	• Thermocouple not located properly at inlet of hopper	• Secure thermocouple probe into coupling at inlet of dryer
Material not drying	• Process and/or auxiliary filter(s) clogged	• Clean filter
	• Incorrect blower rotation	• Check rotation
	• Regeneration heating elements inoperative	• Check electrical connections; replace elements, if necessary
	• Desiccant assembly not rotating	• Check motor electrical connections; replace motor, if necessary; check drive assembly for slippage, adjust
	• Material residence time in hopper too short	• Drying hopper too small for material being processed; replace with larger model
	• Moist room air leaking into dry process air	• Check all hose connections and tighten if required; check hoses for cracks; replace as necessary; check filter covers for tightness, secure
	• Desiccant contamined	• Replace desiccant cartridge
(b) Granulators		
Stalled machine	1. Overloading	• Feed material slower
	2. Worn, damaged or improperly set knives	• Readjust or replace as required
	3. Screen and/or blower chute blockage	• Check to see if line is clogged and check rotation of blower
	4. Drive-belt slippage	• Check tensioning
	5. Loss of power	• Check power supply, electrical hook-up and safety switches
	6. Motor running in reverse	• Check directions of rotation and rewire per diagram, if necessary

(continued)

Table 12.6: *(continued)*

Problem	Possible causes	Suggested solutions
Material overheating	• See items 1, 2, 3, 6	• Same as above
	• Screen too small	• Change to screen with larger diameter holes
Too many fines in material	• Worn, damaged, or improperly set knives	• Readjust or replace as required
Bearing overheating	• Failure to lubricate properly	• Check frequency of lubrication and type of grease used
	• Too much tension on drive belts	• Adjust tensioning
Excessive knife wear	• Highly abrasive material	• Change to higher alloy knives
Knife breakage	• Tramp metal in scrap	• Check scrap material for foreign matter
	• Loose or stretched bolts	• Check bolts and retorque per specification sheet
	• Uneven knife seats	• Inspect and clean seat surfaces
Screen breakage	• Improperly seated	• Check that screen is fitted correctly
Motor won't start	• Power supply failure	• Check main power supply and fuses
	• Overheated motor	• Allow motor to cool; reset starter overloads
	• Starter failure	• Check for burned-out contacts; replace if necessary
	• Inoperative safety switches	• Check that hand guard is secure and all contact points closed; replace if necessary
(c) Conveying Equipment		
Spiral conveyors		
• Excessive wear	• System may have kinks, sags, sharp or compound bends, or contact with sharp surfaces	• Re-position drive-end tube supports or feed hop
	• System may exhibit excessive vibration	• Inspect for proper feeding into inlet (consult supplier if material is bridging)
	• Are abrasive materials being conveyed?	• Consult supplier
• Excessive spiral wear or breakage	• There may not be sufficient clearance at inlet end of conveyor for spiral to expand taking into consideration length and inclination of conveyor, and bulk density of material	• Shorten spiral

(continued)

Table 12.6: *(continued)*

Problem	Possible causes	Suggested solutions
• Low delivery rate	• Material may not be flowing properly into inlet	• Rotate outer tube so that inlet opening is aligned with hopper feed • Install bin vibrator and/or agitator • Adjust spiral length as per manual instructions • Reverse spiral direction if incorrect • Seal openings in system if material is hygroscopic and system is installed in high-humidity environment
Belt conveyors		
• Slipping clutch	• Oil contamination or excessive lubrication	• Clean system, adjust clutch properly
• Improper belt tracking	• Belt not properly tightened when changed or installed • System damaged in delivery	• Check alignment and tighten
• Electrical malfunctions	• Motor may be exposed to excessive heat under molding machine	• Install overload-protection temperature limit switches
• Parts jam up or fall off	• Transition points not long enough • Parts too wide for machine	• Contact supplier • Identify to supplier what's being conveyed before purchase
• Belt speed too slow or too fast	• Improper adjustment of variable-speed drive motor • Damaged in delivery	• Change sprockets • Install variable-speed options or speed adjusting kit • Specify desired speed range to supplier before purchase
Pneumatic parts conveyors		
• Material won't move	• Filters clogged • Filters improperly sized • Blower binded • System improperly sized to suit plant layout	• Inspect filters regularly and frequently; specify to supplier exactly what types of materials are to be conveyed; check vacuum pressure gauges; replace filters; indicate to supplier anticipated future growth plans, if possible; check vacuum seats.
• Blower overheating	• Blower binded • Excessive ambient-temperature exposure	• Check filters; make sure they are properly sized; install temperature limit switches

(continued)

Table 12.6: *(continued)*

Problem	Possible causes	Suggested solutions
• Blower too noisy		• Install muffler; enclose in a well ventilated sound enclosure; place blower in a sound-proofed room
(d) Metering/Proportioning/Feeding Equipment		
Dry-solids metering		
• Weight distortion	• Dust or adhesive dry solids accumulation	• Install a dust-exhaust system or hardware to "tare off" dust accumulation in critical areas
		• Clean belts, trays, augers periodically
• Mechanical and electrical component failure, improper weight signals	• Dust, environmental conditions, materials adhering to underside of belt or other system components	• Evaluate several systems in production trials before purchase • Clean system components periodically
• Belts stretching and mistracking	• Material buildup or proximity of system to moisture	• Inspect regularly; consult supplier; install corrective recalibration devices
Liquids metering		
• Materials won't flow	• Improperly sized system – can't handle highly viscous materials	• Install drum pump; specify to supplier what type of material to be conveyed
		• Make sure tubing is properly sized
Proportioning loaders		
• Sluggish loading, excessive loading time needed to fill hopper	• Clogged filter in either dust collector or pump	• Clean filter (replace if necessary)
	• Material line clogged	• Clean line
	• Vacuum leak in either material line or vacuum line	• Seal lines
	• Hopper lid or hopper receiver not sealed	• Clamp lid or replace hopper seals if necessary
• Excessive dust carryover into the dust collector	• Valves not sealing	• Check for obstruction and proper air pressure
	• Improper feed tube setting, too much air, not enough material creates high velocities	• Adjust feed tube to give highest obtainable vacuum and smoothest flow
	• Excessive fines and dust in material or improper blending	• Consult material supplier

(continued)

Table 12.6: *(continued)*

(e) Chillers/Mold Temperature Controllers

Problem	Possible causes	Suggested solutions
Cooling water lines frozen	• Thermostat set too low (i.e., below 40 °F without antifreeze)	• Check thermostat and reset if necessary to 40 °F or higher
	• Insufficient antifreeze in process cooling water	• Add antifreeze
System shuts down or cools slowly or poorly while refrigerant pressure is low or drooping; bubbles in refrigerant-level sight glass; oily-looking moisture on coolant-circulating tubes or floor nearby	• Refrigerant leak	• Replace refrigerant, plug leak source, clean condenser
System shuts down while refrigerant pressure is high	• Condenser not getting enough cooling water because constricted or blocked by dirt	• Clean condenser
Low cooling-water pressure	• Leak in process-water circulation lines	• Plug leak
	• Empty water-storage tank	• Fill tank
	• Broken pump motor	• Repair or replace (hermetically sealed motor/pump assemblies must usually be replaced)
	• Broken pump seal	• Replace
Slow or inadequate cooling of molds, high water pressure	• Water flow constricted by dirt or mineral scale	• Clean water circulating system • Treat water • Install intermediate heat exchanger for mold-cooling water (optional)
Slow or inadequate mold cooling, low pressure	• Water-circulating line leak	• Repair leak
	• Pump seal failure	• Replace seal or tighten packing gland
Process water heats slowly or insufficiently	• Heating element encrusted with dirt, mineral scale or, in oil-circulating systems, carbonized oil	• Clean or replace heating element, treat water, replace oil

(f) Dehumidifying Dryer Performance

System	Possible causes	Cure
Cannot attain desired air inlet temperature	• Heater failure	• Check process air or after heaters – regeneration heaters play no part in this aspect of operation

(continued)

Table 12.6: *(continued)*

System	Possible causes	Cure
	• Hose leakages and excessive length air inlet side	• Locate and repair; if the hose is old and brittle, replace • Shorten all hose to minimum lengths
	• Line, hopper, filter blockage	• Check for collapsed or pinched lines, valves that are closed (some makes have air flow valves located on the air inlet side of the hopper); filters should be changed or cleaned frequently; a good trial period is every four weeks until experience dictates a shorter or longer period
Dew point as measured at air inlet to the hopper is unacceptable	• Loss of regeneration heaters in one or both beds or line fuses	• These can be checked with a volt meter at the control panel
	• Loss of timer or clock motor switching ability from one head to the other, i.e., continuous operation on only one dessicant bed	• Check clock motor for movement by observing either function indicators or valve-shifting mechanisms (Note: Loss of regeneration heaters may occur if the clock motor or shifting mechanism malfunctions)
	• Desiccant has deteriorated or been contaminated	• Most manufacturers suggest checking the desiccant annually and replacing when it does not meet test criteria; typically two to three years is a reasonable interval, depending upon the severity of service
	• Loss of power to one or both desiccant beds	• During regeneration cycle, exterior of the desiccant bed should be hot to touch; check contacts on relays or printed circuit board for flaws, line fuses if so equipped
Low or nonexistent air flow	• Fan motor burnout	• Replace
	• Loose fan on motor shaft	• Tighten
	• Clogged filter(s)	• Change
	• Restricted or collapsed air lines	• Correct and relieve restrictions
	• Blower motor is reversed	• Use of a pressure gauge or flow meter is suggested; proper rotation is that in which the highest flow is indicated

12.1.1 Rocker Bottoms and Oval Necks

Warpage in blow molded containers is most often seen as rocker bottoms and out-of-round necks. In general, the cause of the problem is the same as found with other molding processes – inadequate or unequal cooling prior to removing from the mold.

There are a number of possible causes to consider. The study should proceed from the simplest to cure to the most complex:

- Insufficient cooling water to the mold. See if an increased flow will stop the warpage.
- Too high a stock temperature. Lower the temperature by small increments to see if it will effect a cure without causing new problems, such as cold spots in the container wall.
- Blocked cooling channels. Check out the throughput, and if found appreciably less than when the mold was new, clean the channels.
- If the above conditions have not achieved a cure, consider whether the cycle is too unrealistically short to keep productivity high. If not, proceed to other possibilities.
- Poorly designed cooling channels. If this is determined to be the case, the mold may have to be reworked to increase cooling capability or perhaps to improve the uniformity of cooling.
- Poorly designed part. Too great a variation in the distribution of material in the part – unnecessarily thick and thin sections – readily cause warping unless that part is thoroughly cooled, which may demand an uneconomically long cycle. Depending upon the situation, parison programming or a redesign of the part itself may be the only feasible solutions.

12.1.2 Defects Within the Blown Wall

Various defects found within the blown wall will be evaluated in this section. They include defects that spoil the appearance of the blown part such as "bubbles" in the wall and "cold spots." In addition, there are 3 distinct types of weld or pinch-off defects including thinning, tearing, and cutting of the pinch-off. Another defect discussed are indented parting lines.

12.1.2.1 Bubbles

Moisture most likely cause

These defects are not easily mistaken for foreign matter in the melt for they show up as generally symmetrical areas, clearer than the surrounding area. Moisture in the melt is the most likely cause. It may come from several sources:

- Cold resin that is brought in and exposed to the warm atmosphere of the extruder room will condense moisture on its surface. If the resin is not allowed to warm up and evaporate the moisture before entering the extruder, the moisture will end up as bubbles in the blown part.
- If water cooling is overdone on the throat, moisture may condense in the barrel, and if running at slow screw speeds, the result will be the same as if the resin were wet.

Nonmoisture causes of bubbles include material leaking from mismatched head sections and insufficient back pressure.

Curing the moisture problem is generally self-evident. Let the resin warm up before using by keeping a few unopened gaylords in a warm room. If drawing directly from hopper cars or trucks, it is advisable to discharge into a surge bin which will protect the resin from moisture condensation and allow it to warm up before use. If condensation is occurring in the barrel, simply ease off on throat cooling.

If mismatched head sections are found, the situation should immediately be rectified, for this can also be a source of degraded resin.

Increasing back pressure, if this is suspected as a cause, is easily accomplished on reciprocating units by adjustment of the back pressure gauge. With continuous extruders, a screen pack change to a finer mesh may solve the problem.

12.1.2.2 Cold Spots

Caused by non-uniform melt

These defects show up as non-homogeneous areas in the wall of a blown part. The cause of this defect is too low a back pressure and insufficient working of the melt to ensure that it is completely uniform. Apply the same remedy proposed for bubbles.

Another cause is accidental admixture of some resin having a lower melt index. Extrusion conditions will be inadequate resulting in cold spots from the incompletely melted resin. The obvious cure is to not let such accidents occur.

12.1.2.3 Thinning

Temperature

A thin, weak weld line is often caused by too high of a stock or mold temperature. As the molds close and the pinch-off is made, the pinch-off lands fail to force sufficient material into the weld line to make a strong weld. Gradually drop the stock temperature to see if this will cure the problem. If it does not, then examine the pinch-offs. Too short or too sharp a pinch-off land length will nearly always result in a thin, weak weld line.

Excessive pressure

Excessive preblow or high pressure air coming on too early can also cause the pinchoff to thin before it has time to set. Gradually back off the preblow air to a point that does not affect the other areas of the container. If this does not correct the problem then the blow delay time can be increased. Care must also be taken when performing this step because other areas of the container may be affected.

12.1.2.4 Tearing

Check pinch-off land

Weld tearing during trimming is usually a result of too long a pinch-off land length. This may have been selected in the first place to achieve a strong weld by forcing more material into the weld line. It will accomplish the latter, but, if overdone, can prevent the molds from closing completely. This leaves a thick pinch-off line that is apt to tear during trimming, leaving a rough, ragged weld, or the weld may even be torn open. Have the molds reworked to reduce the land length to a point that a good balance is achieved between a strong weld and a trimmable pinch-off. A land length of 0.010 to 0.015 in. (0.254 to 0.381 mm) is most commonly used for polyethylene and most other plastics.

Other areas to investigate when tearing becomes a problem are worn pinch-offs, mismatched molds, damaged molds, and uneven mold clamp pressure. Worn pinchoffs can be temporarily corrected by rolling them back, but since this is only temporary, they will need to be refurbished.

Mismatched molds are normally caused by worn locating pins which will need to be replaced. Damaged molds, such as nicks in the pinch-off area, will need to be filled and reground. Molds that appear to have torn pinch-offs in one area and not another are probably not closing and clamping evenly. The clamp will need to be adjusted. If this cannot be accomplished, the molds can be shimmed out at the problem area.

12.1.2.5 Cutting

Check temperature

If the weld and pinch-off both seem to be satisfactory but there is a hole or slit along the weld, the stock temperature may be set too low. If the stock is too cold, the parison may tear as the mold closes. Check a part for cutting as it comes out of the mold with flash attached. If a hole or slit is found, this eliminates the trimming operation as the cause. Examine the mold for defective pinch-offs. If none are found, raise the stock temperature by 5 to 10 °F (3 to 6 °C) increments until the cutting ceases.

Cutting can also result if the molds are allowed to close too fast, slamming or slapping shut. Even more disasterous, this condition may cause the molds to become severely damaged. Ease off on the closing cycle to make better weld and to preserve the molds.

12.1.2.6 Indented Parting Lines

Pressure or venting

Indented part lines have somewhat of a V shape where they appear to pull into the container. This problem is normally caused by insufficient blow pressure or air entrapment due to poor venting. If raising the blow pressure does not cure the problem, the mold vents need to be cleaned. Should the problem continue to exist, the molds will probably require sand blasting to improve venting.

12.1.3 Poor Bottle Surface

Roughness, pits, orange peel, and a variety of other terms describe the poor surfaces frequently found on blown bottles. While an imperfect parison may be a cause, only problems associated with the mold and the blowing process will be evaluated.

Caused by mold or process conditions

- Poor mold surface. The mold should have a fine matte finish in order for the air to vent quickly and allow the parison to conform to the mold surface while still hot.
- Plugged or inadequate vents. If the vents along the parting line are dirty or plugged, the part surface will suffer. These vents will need to be cleaned. Vents that are too small or too few in number to permit the rapid escape of air will also cause surface problems. Minor reworking of the mold is the answer.
- Low blow pressure or blow rate. Increase the blow pressure, then check to see that the blow pin is large enough to handle the required amount of air to fully and rapidly blow the part. Check for restrictions or partial plugging of the air lines.
- Air leak around the blow pin. Leakage here prevents sufficient pressure to be maintained to hold the part tightly against the mold for good conformation to shape and rapid cooling.
- Low stock temperature. If the parison is too cold, it will be difficult to make it reproduce the surface of the mold, particularly if there is lettering or fine designs on the surface.

Gradually increase stock temperature cautiously. Overdoing this may cause problems that result from overly high stock temperatures.

- Condensed water in the mold. A low mold temperature combined with high humidity in the plant can result in water condensing in the mold between the ejection of a part and the blowing of the next. This can cause bad surface pitting. If the mold temperature cannot be raised, usually not without hurting cycle time, it may be necessary to air condition the area locally or in the entire plant to remove some of the water in the air during periods of excessive humidity.

Caused by parison

A particular kind of poor bottle surface results from a rough parison. Such a parison is almost always caused by operating at a shear rate that produced melt instability.

A region of melt instability is a fundamental characteristic of all HDPEs. The region starts at about the upper level of extrusion rates at which continuous extruders are capable, and ends near the lower level of extrusion rates at which reciprocating and accumulator extruders are likely to operate.

Therefore, while running a continuous extruder, if the type of surface roughness indicative of melt instability is encountered, back off on the extrusion rate until the problem disappears.

With either a reciprocating or accumulator type, there are several choices:

- Drop down out of the region of melt instability by further decreasing extrusion rates (increasing drop time), a generally unsatisfactory solution because of the decrease in productivity.
- A better approach is to move up out of the region by increasing extrusion rates (decreasing drop time).
- If it is impractical to increase extrusion rates, try increasing the stock temperature or change to a higher melt index resin.

A second type of rough surface is caused by melt fracture, rarely seen in blow molding, but would result from running at shear rates that only the accumulator or reciprocating type extruders can reach, and only then, if pushed to the limit of their performance. Obviously, the answer is to back off on extrusion pressure until the melt is back in stable territory.

12.1.4 Curtaining and Webbing

Curtaining or folding of the parison as it extrudes from the die is the cause of webbing in necks and handles. If the effect is severe enough, the handle can be partially or completely blocked. Folds in the neck can interfere later with filling operations, or be so restricted as to cause difficultly in reaming and facing, and spoiling the fit of the cap. The several possible causes all imply their own cure, which should be quickly undertaken to eliminate this serious bottle defect.

Causes include:

Tooling

- Poor tooling design. Too short a land length between the die (bushing) and mandrel will result in poor control of flow through the die gap. There will be areas of high and low flow, and folding will result.

- Die misalignment. Die and mandrel misalignment will also produce differential flow rates in different areas, resulting in folding. Die
- High stock temperature. Unduly high stock temperature will yield a parison with low melt strength which will collapse before it can be blown, and will fold. Temperature
- Resin mismatch. If a resin is chosen with too high a weight swell for the job, the attempt is often made to reduce part weight by decreasing the die gap. The result may be a parison so thin that it quickly collapses and folds. A resin with low diameter swell characteristics may cause lower handle webbing and, in severe cases, can even cause blowouts. This is due to the parison being caught in two places by the inner handle pinch-off. When blown, these areas meet to form a heavy area known as a web. Resin
- Low parison diameter swell. High stock temperature and/or low shot pressure will yield a parison with low diameter swell resulting in lower handle webbing.
- Mold position. Improperly positioned molds that are shifted away from the parison will result in containers having lower handle webbing. Mold
- Hooking parison. A parison hooking away from the handle will also cause the problems stated above. Parison

12.1.5 Blowouts

Blowout of a part can be caused by diverse factors; fortunately the cure in most cases is self-evident. Possible causes are listed below:

- Pinch-off too sharp. This will cut the parison and cause a blowout. Increase the width of the pinch-off land.
- Pinch-off too wide. Too wide a land will cause mold separation and the chance of a blowout.
- Clamp pressure too low. This also can result in mold separation and possibly a blowout.
- Pinch-off too hot. Poor or uneven mold cooling can prevent the pinch-off from cooling sufficiently causing it to cut the parison as if too sharp. Improve the cooling by one of the methods previously decribed under "Warping."
- Blow pressure too high. Stretching the parison at too high a rate can easily lead to blowouts. Back off gradually on the pressure.
- Parison too short. This won't be caught at the bottom pinch-off and guarantees a blowout – the remedy is obvious.
- Blow-up ratio too high. Too great a difference between the diameters of the parison and the part will cause the parison to stretch too much. A thin spot will develop and a blowout will occur. A change of head tooling to a larger diameter die is indicated.
- Resin swell too low. This relates to the previous item, and can be analternate cause. Too little swell results in a smaller than expected parison, which can cause a missed handle accompanied by a blowout. A higher swell resin is needed.

12.1.6 Parison Curl, Stringing, Hooking, Sag, and Length Consistency

In this section, various problems with parison formation will be evaluated including curl, stringing, hooking, sag, and length inconsistency.

12.1.6.1 Parison Curl

Parison curl, sometimes called doughnutting because the parison curls about in a seeming attempt to form itself into a doughnut, usually results from one of three causes:

Main causes

- Cold mandrel or die
- Misaligned die and mandrel
- Foreign matter or degraded resin in the die (bushing)

If the curl occurs when a machine is started up, then gradually disappears as the unit approaches operating temperature, a cold mandrel or die can be suspected. Allow a longer warmup time before attempting production. Also, check to see that the die heater, equipped with most blow molders, is operative. If it is not, an unnecessarily long warmup time will be needed.

Parison curl on a fully warmed up machine may result from a misalignment of the die and mandrel. The mandrel edge may be recessed within the die so that the parison contacts the die, hangs up on one side and thus curls. Often, the mandrel has deliberately been so positioned in an attempt to blow a lighter-weight bottle than the tooling was designed for. New tooling should be obtained or the die can be machined to bring the mandrel back to a flush or lower position with the die face.

Another possible cause is an uneven buildup of foreign material on the die, which then distorts the parison as it is extruded. This solution is also self-evident: give the die a thorough cleaning.

12.1.6.2 Parison Stringing

Temperature

Parison stringing is seen as feathery strings of melt attached to the top and bottom flash of the bottle. If stringing is severe, each bottle remains attached to the next. The lower bottle's weight may thin out the parison to the point that blowout can occur, especially in a handled bottle. The stringing is almost always caused by too high a stock temperature and can be eliminated by gradually lowering the temperature.

Fill pressure

Parison stringing can also be caused from too high a fill pressure on reciprocating type extruders, resulting in excessive weeping of polymer from the die. This excessive weeping causes the parison to string during part removal.

In rare cases, when the temperature is lowered enough to eliminate the stringing, other problems, such as cold spots, can arise. In such an event, it may be a matter of the wrong resin being selected for the particular job.

12.1.6.3 Parison Hooking

Parison hooking appears as a parison that will not drop straight, but is hooking from one side or another, causing poor wall distribution, webbing, and/or blowouts in the handle. Parison hooking usually results from any one or a combination of the following causes:

- Nonuniform die temperature
- Warped die or mandrel
- Die gap out of adjustment
- Air flowing on parison

Main causes

12.1.6.4 Parison Sag

Parison sag or drawdown is seen as a parison that thins closest to the die from its own weight during extrusion. Causes are:

- High stock temperature
- Slow extrusion rate
- Long mold open time
- High die heats
- Resin has too high a melt index

Main causes

12.1.6.5 Parison Length Inconsistency

Parison length inconsistency is identified as a parison changing from too short, causing blowouts in bottom of container, to a parison that is too long, causing wall distribution problems and tails sticking to the container during unloading. Parison length can change for any of the following reasons:

- Insufficient back pressure in extruder
- Changing regrind virgin mixture
- Malfunctioning temperature controllers
- Bridged or collared screw
- Worn barrel of screw

Parison length inconsistency from head to head is self-explanatory. Causes are as follows:

- Nonuniform head temperature
- Nonuniform bottle weights
- Manifold chocks out of adjustment
- Head chocks out of adjustment

12.1.7 Foreign Matter in the Melt

Identify and eliminate sources

Foreign matter in the melt – whether from outside contamination of the resin or from particles of degraded resin – discloses its presence in a variety of ways: by die lines or streaking in the parison, by off-colored particles buried in the wall of the molded part, or by holes or "windows" in the blown part. Obviously, the cure is to identify and eliminate the source of the foreign matter.

12.1.7.1 Look for Exterior Sources of Contamination First

Before investigating the more complex problem of degraded resin, first examine closely the material handling procedures and general housekeeping. Are gaylords allowed to stand open to collect dust? Are bags opened and emptied carelessly so as to introduce dirt, lint, and dust into the feed? Are parts with foreign matter in them being thrown into the grinder? Is the regrind being protected with the same care as the virgin resin? A contamination problem can be just that simple.

12.1.7.2 Degraded Resin – Many Sources

Tracing the source of degraded resin particles can involve some detective work. Logically approach the task by proceeding with the easiest cured source to the most difficult. Various clues will suggest the source. A very simplified guide to degradation problems is presented that can be integrated with other information presented in this chapter and others.

Cause	*Solution*
• Improper shutdown and/or startup procedures	
• Overheated material	• Reduce barrel temperatures
	• Reduce residence time, shorten cycle
• Temperature control malfunction	• Calibrate controllers
	• Check controllers
	• Check associated relays, etc.
	• Check thermocouples
• Material hang-up or dead spot in system	• Clean system, purge compound
	• Disassemble and clean
	• Improve material flow by streamlining design
• Contaminated material	• Inspect material to be sure it is clean

If a single streak, or perhaps two or three, suddenly appears in the parison, it is likely that some foreign matter has become lodged between the die and mandrel. Just increase the die gap briefly to purge the particles, then reset to the original gap.

If parison streaking or defective bottles (foreign particles or holes) are regularly appearing and all possibilities of the resin feed being contaminated have been eliminated, then a steady supply of degraded material may be generating somewhere within the head, adaptor, or extruder. The cause will probably be a blind spot or nonstreamlined area where melt hangs up, degrades, and then breaks away from time to time. A poor fit between head and adaptor or adaptor and extruder barrel, for example, can produce a sharp break in the melt path and allow resin hangup. Reworking the parts for a smooth, streamlined fit is the solution.

The gradual appearance of many die lines or streaks, after extended periods of operation, points to a definite buildup of carbon or degraded resin in the head. Fortunately, buildup most often occurs on the die mandrel which are readily accessible for cleaning. Extreme caution should be taken to prevent damaging the machine in the process. Use a tool made of copper or brass which is softer than the machine parts and will not mar or scratch their surfaces.

If the dead spots in the head have been eliminated as the real source, there is a need to go deeper into the problem and find the source of degraded resin somewhere back in the extruder.

12.1.7.3 Culprit – Man or Machine?

Degraded material generated in the extruder can result from defective machinery or poor operating practices. Examination of the extruder head may disclose the source: mismatched or nonstreamlined head sections, or inadequate seals that permit material to leak from supposedly matched surfaces.

Often operating errors

But more often, the degraded resin will result from some operating error. A poor purge when changing from one resin to another may be the cause. A flood of specks in the bottles upon startup, after a prolonged shutdown, will suggest that the extruder was not allowed to cool down on purge before shut off. The resin in the die lips become oxidized by exposure to the air. This cause can be eliminated, and both shutdown and startup times shortened by the use of a simple air-lock plate to seal off the die lips from the atmosphere during these periods.

In other cases, the degradation may be occurring back in the extruder barrel. The cause could simply be running at too high a temperature somewhere or along the barrel. The high temperature may be intentional in an effort to run a difficult resin and still maintain a fast cycle. A slower screw speed will probably solve the problem but will also increase the cycle time. Since this is rarely acceptable, the better answer is to switch to a resin that is easier to extrude.

The high temperature may be unintentional. This could result from a broken thermocouple or malfunctioning temperature controller.

In other cases, too much heat will be generated at the wrong point – in the metering zone. Generally, it is best to let this zone hold its contribution to a single function – metering. Try a reverse temperature profile; rather than gradually increasing the temperature of each zone along the extruder, increase it in the transition section, drop it in the metering zone, then let it smoothly increase to the desired final melt temperature through the adaptor and head.

Once a workable temperature profile has been found, a good purge may clear up the source of the degraded resin. But if the bad condition has existed too long, it may be necessary to take the extruder down for a thorough cleaning.

12.1.8 Die Lines or Streaking in the Parison

Check heater band

Parison streaking most often results from particles of foreign matter or degraded resin lodged between the die and mandrel. Since the die gap is no longer of uniform width, the parison is reduced in thickness at each point where a particle is present. This uneven thickness shows up as streaks or die lines in the parison as well as in the blown part.

There is a type of parison streaking that has nothing to do with foreign matter. This shows up as clear or slightly rippling streaks. In such a case before looking for nonexistent sources of foreign matter, check the heater bands. A burned-out band will cause a cold spot in the barrel, melt will flow unevenly here, dragging or hanging up and cause the streaking. Replacement of such a non-functioning band will immediately eliminate this type of streaking.

12.1.9 Shrinkage

Shrinkage of a polyethylene (or other plastics) container after molding is a normal occurrence. However, excessive shrinkage can be caused by insufficient cooling during the molding cycle or heating of the container during storage.

Excessive shrinkage can be caused by:

Main causes

- Wall thickness or weight too high
- Mold temperature too high
- Stock temperature too high
- Blow pressure too low
- Blow time too short
- Mold volume incorrect
- Bottle storage area too hot
- Resin density too high

12.1.10 Causes of Stress Cracking

Diagnosis of stress crack failures in high density polyethylene bottles is, often, far more complex than just an interaction between the container and its contents. Although specifically for HDPE bottles packaging liquids, many of the analyses can also be applied to similar problems in plastic tote boxes, bowls, or even plastic piping.

For blow-molded containers designed to package a liquid, the ability to withstand the stresses in the presence of a standard liquid surfactant is generally referred to as its environmental stress-crack resistance (ESCR).The value is reported as F50, hrs; that is, the period of exposure time at which there is a 50% chance that the container will fail.

Actually, four basic types of stress cracking are recognized: environmental, solvent, mechanical, and thermal. The last is rarely, if ever, applicable to blow-molded containers. The other three definitely apply and may appear alone or in combination. Although stress cracking is associated with many polymers, researchers were specifically concerned with HDPE, since containers blow-molded from these resins are used for a greater variety of applications and in greater volume than any other polymer.

12.1.10.1 Environmental Stress Cracking

Environmental Stress Cracking (ESC) is the brittle failure of a container filled with a polar liquid at a point of stress concentration while under a stress. All of these factors must be operative.

The stress may be the weight of the container contents or the internal pressure in the container. A point of stress concentration may be a sharp discontinuity in the container design (for example, a sharp change in configuration or a stress point molded in) or caused by differential rates of cooling of thick and thin sections. The polar liquid is one that does not perceptibly swell, chemically attack, or in any other way affect the plastic.

12.1.10.2 Solvent Stress Cracking

This cracking is similar to, and is often confused with, ESC. All factors characteristic of ESC must be present except for liquid content; this one does affect the plastic. It may be absorbed by the polymer, causing it to swell, soften or embrittle, or otherwise exhibit change. There may actually be a chemical attack on the plastic.

12.1.10.3 Mechanical Stress Cracking

This action refers to external stress, typically caused by top loading. It may be this stress rather than an internal stress that contributes to an ESC problem.

Table 12.7: Analyzing Stress Crack Failures in HDPE Bottles

Category	Specifics
Resin	• Homopolymer or copolymer
	• Resin supplier designation, lot number
Corrugated appearance	• Package design
	a. Test weight
	b. Correct dimensions
	c. Divider design – correct dimensions
	• Evidence of damage to package
Bottle history	• Location and date of manufacture
	• Storage conditions
	a. Temperature
	b. Humidities
	c. Stacking height
	d. Failure positions within stack
	e. Date failures discovered
	f. Date reported
	g. Percent of total shipment represented by failures
Product compositions	• Surfactants (wetting agents)
	• Alcohols
	• Other ingredients with high vapor pressure
	• Has product compatibility test been run?
Miscellaneous	• Location of failure on bottle
	a. Bottom chime
	b. Neck
	c. Other
	• Lid seal (tight or venter?)

Source: Norchem/Enron Chemical Co.

When an HDPE container fails in service, it is generally assumed to be an ESC problem. This may be misleading. The resin might not be up to handling the particular product packaged in it. This, however, could have been discovered through a product compatibility test. Such a test should always be made unless the product is essentially identical to others that have been successfully packaged in containers of the same resin.

Where solvents are part of a packaged product, a compatibility test is a must. This need is often overlooked if solvents are only a small part of the formulation, or if certain ingredients are not deemed to be solvents. A petroleum distillate would rarely be overlooked, but others can be, such as alcohols, ketones, aldehydes, chlorinated aromatics and aliphatics, essential oils, etc.

A factor which may set off a futile chase along a false trail is the packaging of the containers or the handling of these packages. By error or by a seemingly harmless change in the specifications for the corrugated shipping container or its divider, the package integrity can be weakened and a greater burden placed on the containers inside. The same is true if the corrugated test strength is decreased or if the height of the case and its dividers are increased.

Similar problems can be caused by cases being stacked higher than recommended or tested. These packaging changes can cause mechanical stress crack failures in the bottles.

Because the causes of stress cracking can be complex, the checklist in Table 12.7 is offered to prevent some apparently minor, but possibly, important factors from being overlooked.

12.1.11 Heat – Too Much or Too Little and What to Do About It

Temperature problems are common

As any blow molder operator knows, temperature problems can occur at any time, affecting quality of the product, bottle weights, cycle times, or combination thereof. More often than not, the problem is a matter of too much heat, but cases of too little heat are not rare. Some suggestions for fast, systematic pinpointing of heating problems are discussed below.

Too much heat – consider the following:

- Wrong set point. Even if one "knows" this can't happen, check it out first. This only takes seconds and may avoid misspending a lot of energy looking for other possible causes.
- Insufficient cooling. If the controller is not calling for heat, yet a zone seems to be overheating, the heating problem may actually be a cooling problem, so check for proper coolant flow.
- *Shorted thermocouple.* Check the controllers to see if the temperature indication is at or near zero. If it is then the most likely cause of the heating problem is a shorted thermocouple. This will deliver no voltage to the indicator, and the controller will call continuously for heat. A broken thermocouple would produce the same result as a shorted thermocouple in very old instruments not equipped with thermocouple break protection.
- *Poor contact between thermocouple and barrel.* If the thermocouple is not making good contact with the metal of the barrel, it will sense a temperature lower than actual and will cause the zone to overheat.
- *Partial short in heater band.* A partial short will reduce the electrical resistance of the heater band and, as a result, will increase the wattage output. This possibility will take a few minutes to check out. First, note the voltage and wattage values stamped on the heater

at the zone believed to be the trouble spot. Then, with the electrical supply turned off, disconnect one end of each of the heater bands in the zone and measure the resistance.

Now, using the voltage and wattage values stamped on each heater band, calculate the resistance, using the following formula:

Calculate resistance

$$\text{Resistance} = \frac{(\text{Voltage})^2}{\text{Watts}}$$

For example, if the voltage reads 220, and the wattage is 2,000, then,

$$\text{Resistance} = \frac{220 \times 220}{2{,}000} = 24.2 \text{ ohms}$$

If the resistance value, as measured, is appreciably lower (10 % or more) than the calculated value, then that heater band should be replaced. The actual voltage should also be measured to be certain it is as it should be. If it is not, the problem is probably in the power pack and not the heater.

Check heater band

Too little heat – consider the following:

- *Wrong set point.* Check it out.
- *Broken thermocouple.* When checking the set point on the controller, note if the indication has been driven to the top of the scale. If so, you very likely have a broken thermocouple, and its replacement should cure the problem.
- *Low voltage at heater band.* If a voltage test at a heater band shows low or no voltage reaching the heater, check out the connections back to the controller. A gap in the circuit will, of course, cut all voltage to the heater band. A poor connection, which can put a high resistance in the circuit, will result in low voltage being delivered to the heater.
- *Break or partial break in heater band.* This will increase the resistance of the band and reduce the wattage. Check out the bands with a volt-ohm meter and make the calculation described previously for partial short in heater band.

12.1.11.1 Instrument Recalibration

It's a good idea to check out the instrumentation at least on a semiannual basis. A controller actually controlling 50 °F (28 °C) higher or lower than it indicates can produce some fairly puzzling situations. So, use a pyrometer to test if actual and indicated temperatures agree. If they differ by more than a few degrees, recalibrate.

Quick check

This test – if not the recalibration – can be done in moments, and probably should be placed high up in priority for checking before going on to tests that take more time and/or require the blow molder to be shut down.

12.1.12 Surging

There are several causes for surging, but the first one to check is the one that will present the most trouble if not quickly dealt with. And that is a bridging or partial blocking of feed in the feed throat or feed zone. A collar or rope of partial melted pellets is formed that impedes the even flow of resin and prevents the screw from maintaining a steady pressure on the melt. This usually results from overheating resin in the feed zone.

Check temperature on rear barrel

To test this possibility, measure the temperature on the rear barrel. If over 350 °F (177 °C), this may be the culprit (for polyethylene resins), so reduce it to a safe level. Also, check the temperature of cooling water at the outlet from the feed throat if the extruder is so equipped. The temperature should not be higher than about 100 to 120 °F (38 to 49 °C).

After the temperatures have been corrected, get rid of the collar by feeding in handle slugs from the trim station or cut up pieces to tail flash.

If this does not work, there is another method. Decrease the feed zone temperature still further, increase the temperature in the transition zone and increase the screw speed. Caution: Do not let the head pressure or the motor load rise too high.

One of these methods usually will eliminate the problem, but if neither does, then the last resort is to shut down, pull the screw, and clean it.

If the feed throat and feed zone temperatures are satisfactory, then check for a possible variation in regrind feed rate. This will make waves that can be transmitted all the way to the die.

Variations

Still another cause of surging relates to variations in screw speed. Check the rpm meter for a cyclic variation. Variations can result from several causes, some of which can only be discovered while the process is shut down. Included are:

- Voltage fluctuations
- Loose belts
- Broken gear teeth or worn gears
- Misadjusted variable speed drives

A final possibility for the cause of surging, although seldom encountered, is a broken or contaminated screen pack.

12.1.13 Sharkskin

At one time or another, there may be bottles coming off the line with a sharkskin inner surface – wide, alternating bands of thick and thin wall sections. What happened? If running with continuous extrusion, the problem results from extruding too fast. Just back off a few RPMs on the screw speed and the problem will disappear. But, if operating with reciprocating extrusion, increase the pressure for a faster drop time.

The explanation for this odd state of affairs lies in the complex behavior of HDPE melt as it is being pushed through the die. In general as pressure is increased, the extrusion rate increases (or drop time decreases).

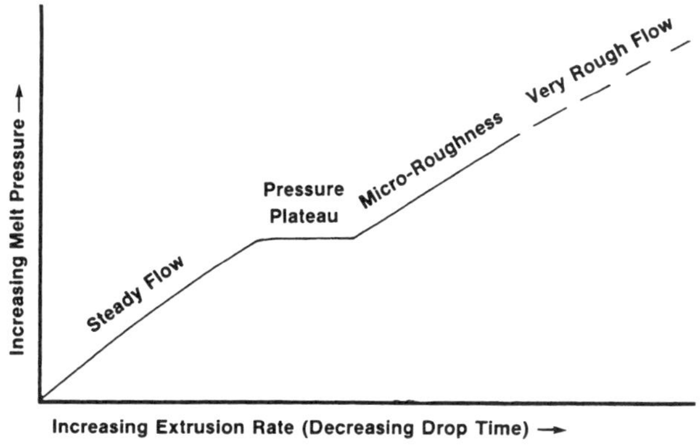

Fig. 12.1: HDPE melt flow in real life

However, when real numbers are plotted, instead of a straight, linear relationship, they form a curve like the one in Fig. 12.1. Starting at some low value, the pressure is gradually increased, and for a time, things go as expected – each increase in pressure causes a corresponding increase in extrusion rate. Melt flow is steady, and bottles are smooth. But at some point, the next small increase in pressure results in instant sharkskin. This occurs at the pressure plateau in the curve.

By increasing the extrusion rate to a certain point, the sharkskin vanishes. Actually, neither the parison nor the bottle are as smooth as in the steady-flow region. Most continuous extrusion takes place in the region of steady flow. Sharkskin will be encountered only if the machine is run at extremely high rates. Most reciprocating extrusion takes place above the pressure plateau and sharkskin appears only if extrusion pressure falls too low.

Strange things happen at the pressure plateau; Fig. 12.2 will help explain them. At some point a very slight increase in pressure causes the extrusion rate to almost double. With this rapid rate increase, the pressure drops slightly, and the extrusion rate falls considerably. Then, with

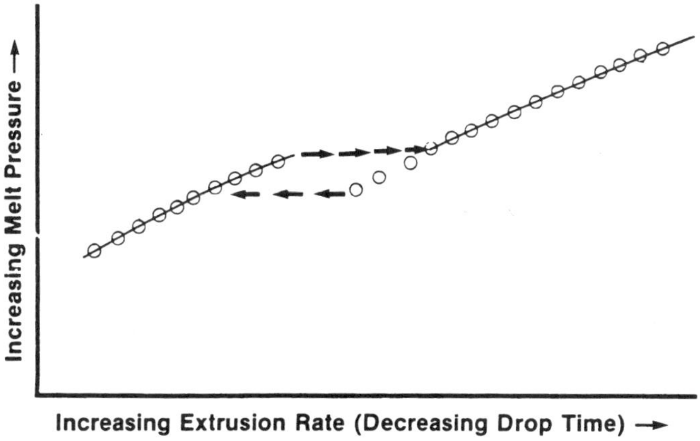

Fig. 12.2: Pressure and extrusion rate oscillations in the pressure plateau regions produce sharkskin

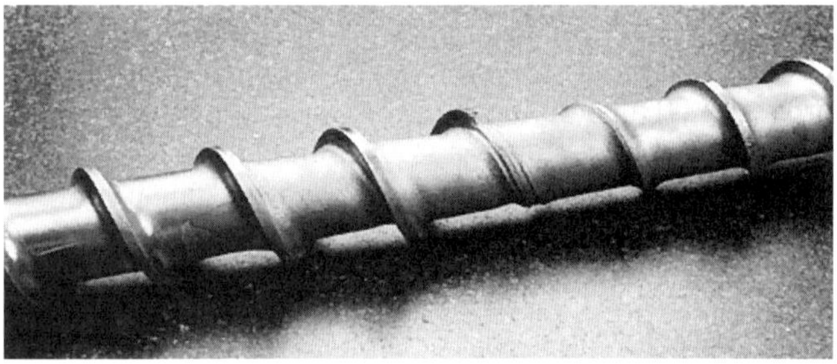

Fig. 12.3: Abrasive wear on screws (and barrels) is caused by abrasive fillers such as calcium carbonate, talc, glass fibers, titanium dioxide, etc. The mechanism of abrasive wear is much the same as emery cloth on metal such as the wear on these screw slights

this decrease in extrusion rate, the pressure builds again, and the cycle repeats, each cycle producing a pair of light and heavy sections in the parison, and in the bottle. So, to stay off the pressure plateau, it is only necessary to reduce or increase the melt pressure.

An unexpected feature of the plateau is that the critical pressure corresponding to the upper limit is basically the same for all high density polyethylenes. There is, however, some control on the *width* of the plateau.

As an example, Enron/Norchem's milk bottle resin 6009, because of its high shear sensitivity, produces a narrower plateau than resins with poor shear sensitivity. The reason is that less pressure is required for a given extrusion rate, and so the onset of the plateau is delayed and its width is less.

Thus, extrusion rates can be increased for better productivity. And for reciprocating extrusion, drop times can be decreased further before the pressure is applied that results in rough parisons and unusable bottles.

Fig. 12.4: Inspection rollers make it easy to inspect screws that otherwise would be difficult to handle because of their weight and length

12.1.14 Screw Wear Affects Performance

One of the less commonly discussed aspects of maintaining injection and extrusion equipment is the inspection of screws (Fig. 12.3) and barrels. From time to time they should be examined to determine whether they are in condition to render the services expected. Screws do not have a continuous outside diameter. This requires special techniques in manufacturing and inspection. The following methods give reliable results and save time.

12.1.14.1 Inspection Rollers

Proper visual inspection of a screw or barrel requires that it be turned many times in order to see all sides. Screws and barrels are often heavy and difficult to turn when supported by the usual means. Fig. 12.4 shows a roller support device that uses two sets of double conveyor rollers supported by multi-slotted angle irons. These angle irons are mounted on plain wood blocks that can be spaced to accept the screw.

12.1.14.2 Diameters

Measure OD

The shank and many other diameters are easy to measure by the usual methods. Other diameters, such as the root diameter or the outside flighted diameter, require special methods. Measuring the root diameter is not always a reliable way to obtain channel depths. Another problem: if the micrometer sits on the radius on both sides, it can give a false reading. If the outside diameter (OD) is severely worn, this is still the best method to determine the correct channel depths. Pitches less than square or very deep channels make the problem worse. The best way to obtain root diameter is to find the OD, then subtract the channel depths.

The OD is measured with the assistance of a "mike" bar spanning two flights (Fig. 12.5). The thickness of the bar is subtracted from the measurement obtained. The usual technique is to place the bar on top and hold the anvil of the micrometer against the slight at the bottom with the left hand. The right hand adjusts the micrometer while making rocking motions along the screw. The bar will rock with the micrometer as the final setting is reached. Of course, it is essential that the bar be straight and of uniform thickness. It is best to check the OD at 90° from the original set of measurements because screws can be manufactured egg-shaped or become worn that way.

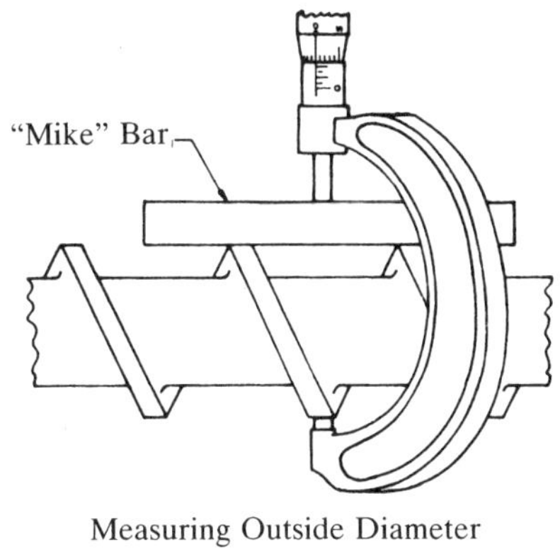

Fig. 12.5: O.D. measurement

Fig. 12.6: Channel-depth gauge

12.1.14.3 Depths

Check channel depth

The channel depth of small screws can be easily checked with a standard depth micrometer. With larger screws, the depth micrometer will not span from flight to flight. If checking many screws, it is best to make a screw-depth indicator or buy one. The screw-depth indicator consists of a wide angle "V" block with a dial indicator mounted on top and the probe extending down through the center of the V. The indicator is placed with the V resting on top of the screw and the probe on top of the flight. The gauge is then adjusted to zero. When the tip moves down into the root, the dial gives an accurate indication of channel depth. It is also very fast, allowing many measurements to be made very rapidly, and it can give continuous readings as the screw is rotated.

This last feature is also helpful in locating the starting and ending points of the feed, transition, and metering sections of a screw. This is not easy to do without the inspection rollers. The original channel depths of a severely worn screw are difficult to determine by this method. In the case of a severely worn screw, it is best to use the root-diameter method. Sometimes deep-jawed calipers can help if micrometers are running on the radii. A Spirex channel-depth gauge is shown in Fig. 12.6.

Fig. 12.7: To obtain a pilot runout measurement at the discharge end of a screw, a sensitive dial indicator as shown here is recommended

12.1.14.4 Concentricity and Straightness

Checking for straightness is difficult for the average plastics processor. If a good, long granite inspection table can be found, it will be helpful in checking concentricity and straightness. A preliminary check for straightness is possible just by rolling the screw on the table. If the screw is not straight, it will roll unevenly and show light under the flights in the low areas. This technique is appropriate only if the screw is not worn. The approximate amount that the screw is bent can be determined by feeler gauges. This technique is not completely accurate because the weight of the screw will tend to straighten it against the table. Most injection screws can be mounted between centers on a lathe and checked with an indicator while they are rotated. This requires an accurate center on both ends.

Extrusion screws usually do not have a center on the discharge end, requiring that a center be installed, then rewelded after testing and possible straightening. Checking the runout on the flighted portion is done with a "T" bar that spans at least three flights. The bar leans against the flights and an indicator measures the movement of the bar.

Straightness and concentricity can be determined on screws that don't have centers by rotating them in V blocks on an accurate inspection table. This is done with the help of a height gauge. This method is particularly useful in inspecting the pilot at the discharge end of injection screws. A sensitive dial indicator is best used as shown in Fig. 12.7.

12.1.14.5 Hardness

The hardness of most portions of a screw is difficult to measure, because the screw is usually too large to test in a Rockwell-type tester. Also, the curved surfaces of the screw present a problem, and it is undesirable to make penetration marks on the surface. A satisfactory

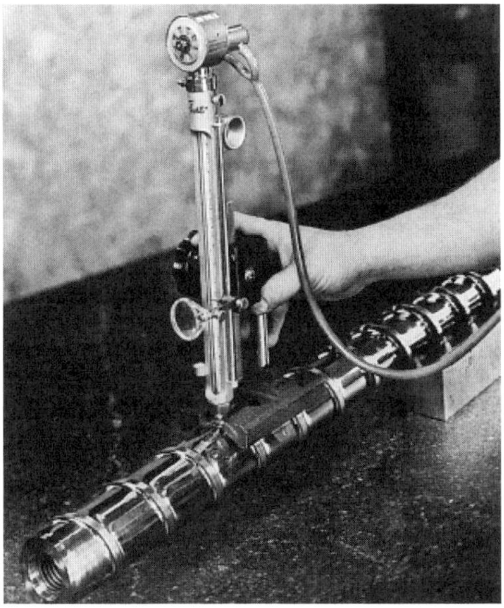

Fig. 12.8: A Shore scleroscope checks screw hardness using the impact or falling ball technique. The Shore device is portable and works on curved surfaces

Fig. 12.9: This portable thickness tester can be used to test chrome thickness or, in the hands of an expert user, check nickel or other coatings

method for checking screw hardness is the impact or fallingball type of tester such as a Shore scleroscope (Fig. 12.8). It is portable, works on curved surfaces, and can be reliable if checked against calibrated reference samples.

12.1.14.6 Finish and Coating Thickness

Finishes can be verified by a number of profilometers. A portable thickness tester can be used to test for chrome-plating thickness (Fig. 12.9), or, with experience, nickel and other nonmagnetic coatings. A surface profilometer is shown in Fig. 12.10.

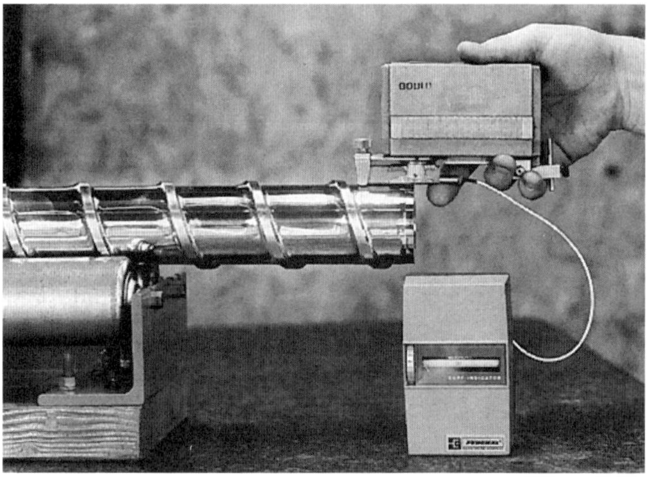

Fig. 12.10: Finishes can be verified by a profilometer

12.1.14.7 Screw Manufacturing Tolerances

For reference, standard screw manufacturing tolerances are published by suppliers and made available on request. These tolerances have been established according to practical application requirements and reasonable ease of machining. Some tolerances can be held closer than indicated on the published list if necessary for a specific application.

12.1.15 Barrel Inspection

12.1.15.1 Inside Diameters

The very front of a barrel can be measured with inside micrometers; however, this is a limited measurement, and will not show far enough into the barrel. An inexpensive cylinder gauge can be rigged up with a long handle to slide the full length. Such a device is accurate enough to determine the need for a replacement or repairs. It is not accurate enough for machining purposes. There is a problem with light and the return glare off the gauge from a flashlight. There are cylinder and bore gauges that measure accurately at great depths and have the indicator outside the barrel for easy reading (Figs. 12.11 and 12.12).

12.1.15.2 Straightness and Concentricity

Barrel straightness is difficult to determine by conventional methods. The internal diameter (ID) and OD are not always exactly concentric. The first method to measure straightness is to set the barrel on precision rollers at each end. The ID is then indicated for runout with a dial indicator. This is limited in depth. Tolerances allow straightness deviations to accumulate to a total allowable indicated runout.

The second method uses a test bar usually about 70% of the barrel length. The test bar is a slotted and chrome-plated bar that is precision-ground to approximately mid-range of the

Fig. 12.11: The Starrett cylinder gauge, available in various sizes, measures accurately at great depths and has an indicator outside the barrel for easy reading

Fig. 12.12: A boregauge, such as this Sunnen Product C. gauge, measures accurately at great depths and permits easy reading

normal screw size tolerances. In theory, if the test bar slides easily through the barrel, the screw will also. In addition, this procedure will catch rapid changes or kinks that would, otherwise, be allowed under the accumulation of tolerances. A different-size bar is needed to test each ID. Because these bars are quite expensive, their use is impractical for most organizations except for the barrel manufacturers.

12.1.15.3 Barrel Hardness

Obviously, the standard hardness testers are not able to get inside a barrel to measure hardness. The instrument most commonly used for this purpose is the internal mobile hardness tester. When testing and comparing the hardness of bimetallic lines, one must remember that absolute hardness, as measured, is not necessarily in direct proportion to wear resistance. Sometimes one may be measuring the softer matrix rather than the wear-resistant carbides. The degree of lubricity or how well the screw and barrel slide against each other is very important. This is not always related to hardness, but is very critical to barrel and screw wear.

12.1.15.4 Barrel Specifications

Chemistry and hardness information for various types of domestic barrels will be supplied for the processor's reference on request. Chemical information supplied is for the "as cast" condition. The actual chemistry may vary widely after final machining is complete. It is important to note that the chemistry and hardness are not necessarily indicative of wear resistance. Also important is how materials (such as tungsten and carbide) are combined and how near to the bore they are located.

Putting into practice these inspection tips to detect problems and taking steps to solve those problems will enhance the operation of any processor using barrels and screws.

12.1.16 Detecting Container Leakage

It is often thought that a container produced as fast and inexpensively as a blow molded product does not warrant leak testing. Unless a major problem occurs, the incidence of leakage is very small. However, because of the speed at which the containers are produced and filled, such a small incidence of leakers can rapidly multiply into a substantial loss. It is worth noting that a machine producing 1,800 1-gallon milk containers per hour, and operating at 99% efficiency (one leaker in 100 containers) will cost the dairy some $200,000 per year in lost milk alone.

From a leak detection aspect, it is vitally necessary to establish a definition of terms. For instance, what is leakage? It is generally understood that the term implies the passage of commodity through a barrier which is meant to retain it: however, when the search for leakage was started it was discovered that the commodity could penetrate the barrier in several different ways, and required different techniques to find it.

Therefore the air monitor grouped types of leakage into four categories, the first being termed an "open" leak, which means that the barrier is thin and presents no restriction to a commodity passing through an orifice in it. This means that the commodity would experience an immediate pressure drop from one side of the barrier to the other.

The same size of orifice, occurring in a thicker barrier, would present a restriction to the flow of the commodity, resulting in a much slower pressure drop.

This restriction would be even more apparent if the passage had a bend or turn in it. To distinguish this type of leak from the "open" leak, it is referred to as a "channel." The two types present entirely different problems when detecting the leakage.

When there are a number of orifices in a thin barrier or, in other words, more than one open leak, it is referred to as a porous condition. This is not the general concept of porosity, but to distinguish between this and a number of "channel" orifices, the term "seepage" is used. The reason for this distinction becomes apparent when considering the methods available for detecting such leaks on a production basis.

The methods for detecting leaks of blow molded containers, fall into three basic categories:

- *The Achieved Pressure Test* The Achieved Pressure Test consists of filling a container with pressure at a controlled rate of flow and registering, by means of a pressure sensor, that the required pressure has been achieved in the container.

  Leakage from the container will prevent pressure building up to the required level, and therefore, the signal to accept the container will not be initiated.

  It will be seen that the system would be infallible if the orifice through the flow control could be the same size as the smallest leak to be detected. It would only allow pressure to flow into the container at the same rate as the leak would release it, and therefore, no pressure buildup would occur. Unfortunately, feeding the pressure through such a small orifice would take a long time to fill a container, far too long for production purposes. The process is speeded up by using a larger input flow orifice and relying upon the ability of the pressure sensor to distinguish the difference in the pressure buildup between leaking and nonleaking containers. A time limitation for the buildup can also be imposed, often by using the duration of the machine cycle for the test.

Because the size of the feed orifice is directly proportional to the pressure differential caused by leakage, the volume of a container tested by this method is limited by the time allowed for the test. This method can only be for "open" type leakage.

- *The Pressure Drop Test* The second method of leak detection consists of filling a container to a prescribed pressure, then shutting off the flow and determining that the container retains this pressure.

  This test is considerably more definitive than the achieved pressure test and can be used for any type of leakage. It will take longer to accomplish than the achieved pressure test when small volumes are involved, as the pressure must be allowed to "settle" before measurement is commenced. This is necessary as the inrush of air will cause an initial surge of pressure, which will drop substantially immediately when the flow is shut off. Such pressure drop will indicate leakage.

  In the case of larger containers the settling time may well be less than the time taken to fill the container for an achieved pressure test.

- *The Displacement Test* The displacement test can be applied as an achieved pressure or pressure drop test, the difference being that a definite quality of air is released into a container to achieve the test pressure, required. This is accomplished by filling a known volume to a prescribed pressure and then releasing the pressure/volume combination into the container being tested. The main advantage of this method is that it controls the inrush pressure surge and permits a faster test. In some cases, the displacement is actually used to generate the test pressure by using the volume chamber as a pump, which is actuated by contacting the container.

  This test is extremely susceptible to variations in volume and is very useful where such changes in volume are the only indication of leakage. It can, for instance, be used to determine that a cap is sealing correctly by covering the cap area, or the entire container if necessary, and releasing the known pressure/volume combination into the cavity created by the sealing action.

  Should some of the quantity of air introduced into the cavity escape into the container, via a leaking cap, the pressure build-up will not be achieved, or hold, as the case might be. This is true even though the cavity left in the container, when filled with a product, is minute.

These three methods do not, by any means, represent the only methods of leak testing available, but they are the methods most used in the testing of blow molded products.

Combination of the methods can be employed, such as controlling the flow into a container for a pressure drop test. This helps to prevent the inrush surge, and automatically includes an achieved pressure test. By using a timer to shut off the input flow, the combination of controlled pressure and time becomes a form of displacement test.

Appendix

Plastics Acronyms

In the plastics industry it is common to define a polymer by the chemical family it belongs to and assign an abbreviation based on the chemistry. However, many times instead of using the standardized descriptive symbol, engineers often use the tradename given by the resin supplier.

This book uses the standardized notation presented in Table C 1. The symbols which have been marked with an asterisk (*) have been designated by the ISO standards, in conjunction with the material data bank CAMPUS. Furthermore, the acronyms presented in Table C 1 may have additional symbols separated with a hyphen, such PE-LD for low density polyethylene, or PVC-P for plasticized PVC. The symbols for the most common characteristics are presented in Table C 2.

Table C 3 presents the most commonly used plasticizers and the symbols used to describe them.

Table C 1: Alphabetical overview of commonly used acronyms for plastics

Acronym	Chemical notation
ABS*	Acrylonitrile-butadiene-styrene
ACM	Acrylate rubber, (AEM, ANM)
ACS	Acrylonitrile-chlorinated polyethylene-styrene
AECM	Acrylic ester-ethylene rubber
AEM	Acrylate ethylene polymethylene rubber
AES	Acrylonitrile ethylene propylene diene styrene
AFMU	Nitroso rubber
AMMA	Acrylonitrile methylmethacrylate
ANBA	Acrylonitrile butadiene acrylate
ANMA	Acrylonitrile methacrylate
APE-CS	see ACS
ASA*	Acrylonitile styrene acrylic ester
AU	Polyesterurethane rubber
BIIR	Bromobutyl rubber
BR	Butadiene rubber
CA	Cellulose acetate
CAB	Cellulose acetobutyrate
CAP	Cellulose acetopropionate

(continued)

Table C 1: *(continued)*

Acronym	Chemical notation
CF	Cresol formaldehyde
CH	Hydrated cellulose, cellulose film
CIIR	Chloro butyl rubber
CM	Chlorinated polyethylene rubber
CMC	Carboxymethylcellulose
CN	Cellulose nitrate, Celluloid
CO	Epichlorhydrine rubber
COC*	Cyclopolyolefine-Copolymers
COP	COC-Copolymer
CP	Cellulose propionate
CR	Chloroprene rubber
CSF	Casein formaldehyde, artificial horn
CSM	Chlorosulfonated polyethylene rubber
CTA	Cellulose triacetate
DPC	Diphenylene polycarbonate
E/P*	Ethylene-propylene
EAM	Ethylene vinylacetate rubber
EAMA	Ethylene acrylic acid ester-maleic acid anhydride-copolymer
EB	Ethylene butane
EBA	Ethylene butylacrylate
EC	Ethylcellulose
ECB	Ethylene copolymer bitumen-blend
ECO	Epichlorohydrine rubber
ECTFE	Ethylene chlorotrifluoroethylene
EEAK	Ethylene ethylacrylate copolymer
EIM	Ionomer Copolymer
EMA	Ethylene methacrylic acid ester copolymer
EP*	Epoxy Resin
EP(D)M	see EPDM
EPDM	Ethylene propylene diene rubber
EPM	Ethylene propylene rubber
ET	Polyethylene oxide tetrasulfide rubber
ETER	Epichlorohydrin ethylene oxid rubber (terpolymer)
ETFE	Ethylene tetrafluoroethylene copolymer
EU	Polyetherurethane rubber
EVAC*	Ethylene vinylacetate

(continued)

Table C 1: *(continued)*

Acronym	Chemical notation
EVAL	Ethylene vinylalcohol, old acronym EVOH
FA	Furfurylalcohol resin
FEP	Polyfluoroethylene propylene
FF	Furan formaldehyde
FFKM	Perfluoro rubber
FKM	Fluoro rubber
FPM	Propylene tetrafluoroethylene rubber
FZ	Phosphazene rubber with fluoroalkyl- or fluoroxyalkyl groups
HIIR	Halogenated butyl rubber
HNBR	Hydrated NBR rubber
ICP	Intrinsically conductive polymers
IIR	Butyl rubber (CIIR, BIIR)
IR	Isoprene rubber
IRS	Styrene isoprene rubber
LCP*	Liquid crystal polymer
LSR	Liquid silicone rubber
MABS*	Methylmethacrylate acrylonitrile butadiene styrene
MBS*	Methacrylate butadiene styrene
MC	Methylcellulose (cellulose derivate)
MF*	Melamine formaldehyde
MFA	Tetrafluoroethylene perfluoromethyl vinyl ether copolymer
MFQ	Methylfluoro silicone rubber
MMAEML	Methylmethacrylate-exo-methylene lactone
MPF*	Melamine phenolic formaldehyde
MPQ	Methylphenylene silicone rubber
MQ	Polydimethylsilicone rubber
MS	see PMS
MUF	Melamine urea formaldehyde
MUPF	Melamine urea phenolic formaldehyde
MVFQ	Fluoro silicone rubber
NBR	Acrylonitrile butadiene rubber
NCR	Acrylonitrile chloroprene rubber
NR	Natural rubber
PA	Polyamide
PA11*	Polyamide from aminoundecanoic acid
PA12*	Polyamide from dodecanoic acid

(continued)

Table C 1: *(continued)*

Acronym	Chemical notation
PA46*	Polyamide from polytetramethylene adipic acid
PA6*	Polyamide from ε-caprolactam
PA610*	Polyamide from hexamethylene diamine sebatic acid
PA612*	Polyamide from hexamethylene diamine dodecanoic acid
PA66*	Polyamide from Hexamethylene diamine adipic acid
PA69*	Polyamide from hexamethylene diamine acelaic acid
PAA	Polyacrylic acid ester
PAC	Polyacetylene
PAE	Polyarylether
PAEK*	Polyarylether ketone
PAI	Polyamidimide
PAMI	Polyaminobismaleinimide
PAN*	Polyacrylonitrile
PANI	Polyaniline, polyphenylene amine
PAR	Polyarylate
PARA	Polyarylamide
PARI	Polyarylimide
PB	Polybutene
PBA	Polybutylacrylate
PBI	Polybenzimidazole
PBMI	Polybismaleinimide
PBN	Polybutylene naphthalate
PBO	Polyoxadiabenzimidazole
PBT*	Polybutylene terephthalate
PC*	Polycarbonate (from bisphenol-A)
PCPO	Poly-3,3-*bis*-chloromethylpropylene oxide
PCTFE	Polychlorotrifluoro ethylene
PDAP	Polydiallylphthalate resin
PDCPD	Polydicyclopentadiene
PE*	Polyethylene
PE-HD	Polyethylene-high density
PE-HMW	Polyethylene-high molecular weight
PE-LD	Polyethylene-low density
PE-LLD	Polyethylene-linear low density
PE-MD	Polyethylene medium density
PE-UHMW	Polyethylene-ultra high molecular weight

(continued)

Table C 1: *(continued)*

Acronym	Chemical notation
PE-ULD	Polyethylene-ultra low density
PE-VLD	Polyethylene-very low density
PE-X	Polyethylene, crosslinked
PEA	Polyesteramide
PEDT	Polyethylenedioxythiophene
PEEEK	Polyetheretheretherketone
PEEK	Polyetheretherketone
PEEKEK	Polyetheretherketoneetherketone
PEEKK	Polyetheretherketoneketone
PEI*	Polyetherimide
PEK	Polyetherketone
PEKEEK	Polyetherketoneetheretherketone
PEKK	Polyetherketoneketone
PEN*	Polyethylenenaphthalate
PEOX	Polyethylene oxide
PESI	Polyesterimide
PES*	Polyethersulfone
PET*	Polyethylene terephthalate
PET-G*	Polyethylene terephthalate, glycol modified
PF*	Phenolic formaldehyde resin
PFMT	Polyperfluorotrimethyltriazine rubber
PFU	Polyfuran
PHA	Polyhydroxyalkanoate
PHB	Polyhydroxybutyrate
PHFP	Polyhexafluoropropylene
PI*	Polyimide
PIB	Polyisobutylene
PISO	Polyimidsulfone
PK*	Polyketone
PLA	Polylactide
PMA	Polymethylacrylate
PMI	Polymethacrylimide
PMMA*	Polymethylmethacrylate
PMMI	Polymethacrylmethylimide
PMP	Poly-4-methylpentene-1
PMPI	Poly-*m*-phenylene-isophthalamide

(continued)

Table C 1: *(continued)*

Acronym	Chemical notation
PMS	Poly-α-methylstyrene
PNF	Fluoro-phosphazene rubber
PNR	Polynorbornene rubber
PO	Polypropylene oxide rubber
PO	General notation for polyolefins, polyolefin derivates
POM*	Polyoxymethylene (polyacetal resin, polyformaldehyde)
PP*	Polypropylene
PPA	Polyphthalamide
PPB	Polyphenylenebutadiene
PPC	Polyphthalate carbonate
PPE*	Polyphenylene ether, old notation PPO
PPI	Polydiphenyloxide pyromellitimide
PPMS	Poly-*para*-methylstyrene
PPOX	Polypropylene oxide
PPP	Poly-*para*-phenylene
PPQ	Polyphenylchinoxaline
PPS*	Polyphenylene sulfide
PPSU*	Polyphenylene sulfone
PPTA	Poly-*p*-phenyleneterephthalamide
PPV	Polyphenylene vinylene
PPY	Polypyrrol
PPYR	Polyparapyridine
PPYV	Polyparapyridine vinylene
PS*	Polystyrene
PSAC	Polysaccharide, starch
PSIOA	Polysilicooxoaluminate
PSS	Polystyrenesulfonate
PSU*	Polysulfone
PT	Polythiophene
PTFE*	Polytetrafluoroethylene
PTHF	Polytetrahydrofuran
PTT	Polytrimethyleneterephthalate
PUR*	Polyurethane
PVAC	Polyvinylacetate
PVAL	Polyvinylalcohol
PVB	Polyvinyl butyral

(continued)

Table C 1: *(continued)*

Acronym	Chemical notation
PVBE	Polyvinyl isobutylether
PVC*	Polyvinyl chloride
PVC/EVA	Polyvinyl chloride-ethylene vinylacetate
PVDC*	Polyvinylidene chloride
PVDF	Polyvinylidene fluoride
PVF	Polyvinyl fluoride
PVFM	Polyvinyl formal
PVK	Polyvinyl carbazole
PVME	Polyvinyl methylether
PVMQ	Polymethylsiloxane phenyl vinyl rubber
PVP	Polyvinyl pyrrolidone
PVZH	Polyvinyl cyclohexane
PZ	Phosphazene rubber with phenoxy groups
RF	Resorcin formaldehyde resin
SAN*	Styrene acrylonitrile
SB*	Styrene butadiene
SBMMA	Styrene butadiene methylmethacrylate
SBR	Styrene butadiene rubber
SBS	Styrene butadiene styrene
SCR	Styrene chloroprene rubber
SEBS	Styrene ethene butene styrene
SEPS	Styrene ethene propene styrene
SEPDM	Styrene ethylene propylene diene rubber
SI	Silicone, Silicone resin
SIMA	Styrene isoprene maleic acid anhydride
SIR	Styrene isoprene rubber
SIS	Styrene isoprene styrene block copolymer
SMAB	Styrene maleic acid anhydride butadiene
SMAH*	Styrene maleic acid anhydride
SP	Aromatic (saturated) polyester
TCF	Thiocarbonyldifluoride copolymer rubber
TFEHFPVDF	Tetrafluoroethylene hexafluoropropylene vinylidene fluor
TFEP	Tetrafluoroethylene hexafluoropropylene
TM	Thioplastics
TOR	Polyoctenamer
TPA*	Thermoplastic elastomers based on polyamide

(continued)

Table C 1: *(continued)*

Acronym	Chemical notation
TPC*	Thermoplastic elastomers based on copolyester
TPE	Thermoplastic elastomers
TPE-A	see TPA
TPE-C	see TPC
TPE-O	see TPO
TPE-S	see TPS
TPE-U	see TPU
TPE-V	see TPV
TPO*	Thermoplastic elastomers based on olefins
TPS*	Thermoplastic elastomers based on styrene
TPU*	Thermoplastic elastomers based on polyurethane
TPV*	Thermoplastic elastomers based on crosslinked rubber
TPZ*	Other thermoplastic elastomers
UF	Urea formaldehyde resin
UP*	Unsaturated polyester resin
VCE	Vinylchloride ethylene
VCEMAK	Vinylchloride ethylene ethylmethacrylate
VCEVAC	Vinylchloride ethylene vinylacetate
VCMAAN	Vinylchloride maleic acid anhydride acrylonitrile
VCMAH	Vinylchloride maleic acid anhydride
VCMAI	Vinylchloride maleinimide
VCMAK	Vinylchloride methacrylate
VCMMA	Vinylchloride methylmethacrylate
VCOAK	Vinylchloride octylacrylate
VCPAEAN	Vinylchloride acrylate rubber acrylonitrile
VCPE-C	Vinylchloride-chlorinated ethylene
VCVAC	Vinylchloride vinylacetate
VCVDC	Vinylchloride vinylidenechloride
VCVDCAN	Vinylchloride vinylidenechloride acrylonitrile
VDFHFP	Vinylidenechloride hexafluoropropylene
VF	Vulcanized fiber
VMQ	Polymethylsiloxane vinyl rubber
VU	Vinylesterurethane
XBR	Butadiene rubber, containing carboxylic groups
XCR	Chloroprene rubber, containing carboxylic groups
XF	Xylenol formaldehyde resin

(continued)

Table C 1: *(continued)*

Acronym	Chemical notation
XNBR	Acrylonitrile butadiene rubber, containing carboxylic groups
XSBR	Styrene butadiene rubber, containing carboxylic groups

Table C 2: Commonly Used Symbols Describing Polymer Characteristics

Symbol	Material characteristic
A	Amorphous
B	Block-copolymer
BO	Biaxially oriented
C	Chlorinated
CO	Copolymer
E	Expanded (foamed)
G	Grafted
H	Homopolymer
HC	Highly crystalline
HD	High density
HI	High impact
HMW	High molecular weight
I	Impact
LD	Low density
LLD	Linear low density
(M)	Metallocene catalyzed
MD	Medium density
O	Oriented
P	Plasticized
R	Randomly polymerized
U	Unplasticized
UHMW	Ultra high molecular weight
ULD	Ultra low density
VLD	Very low density
X	Cross-linked
XA	Peroxide cross-linked
XC	Electrically cross-linked

Table C 3: Commonly Used Plasticizers and Their Acronyms

Acronym	Chemical notation
DODP	Dioctyldecylphthalate
ASE	Alkylsulfone acid ester
BBP	Benzylbutylphthalate
DBA	Dibutyladipate
DBP	Dibutylphthalate
DBS	Dibutylsebacate
DCHP	Dicyclohexylphthalate
DEP	Diethylphthalate
DHXP	Dihexylphthalate
DIBP	Diisobutylphthalate
DIDP	Diisodecylphthalate
DINA	Diisononyladipate
DMP	Dimethylphthalate
DMS	Dimethylsebazate
DNA	Dinonyladipate
DNODP	Di-*n*-octyl-*n*-decylphthalate
DNOP	Di-*n*-octylphthalate
DNP	Dinonylphthalate
DOA (DEHA)	Dioctyladipate, also diethylhexyladipate, DEHA no longer used
DOP, DEHP, DODP	Dioctylphthalate, dioctyldecylphthalate
DOS	Dioctylsebacate
DOZ	Dioctylazelate
DPCF	Diphenylkresylphosphate
DPOF	Diphenyloctylphosphate
DPP	Dipropylphthalate
ELO	Epoxidized linseed oil
ESO	Epoxidized soy bean oil
ODA	Octyldecyladipate
ODP	Octyldecylphthalate
PO	Paraffin oil
TBP	Tributylphosphate
TCEF	Trichlorethylphosphate
TCF	Trikresylphosphate
TIOTM	Triisooctyltrimellitate
TOF	Trioctylphosphate
TPP	Triphenylphosphate

Index

C

D